Birkhäuser

Lecture Notes in Geosystems Mathematics and Computing

The *Lecture Notes in Geosystems Mathematics and Computing* series showcases topics of current interest that lie at the interface of the geosciences, mathematics, and observational and computational sciences. Titles may present new results or novel perspectives on existing areas of research, or serve as extended surveys of certain topics. Traditional lecture notes covering core concepts for graduate and doctoral students are also included, as are titles that are based on expansions of outstanding PhD theses. Topically, titles address either mathematical and computational concepts and techniques – such as inverse problems, applied harmonic analysis, numerical simulation, and data analysis – or focus on practical applications in such areas as gravitation, climate modeling, seismology, geomechanics/dynamics, and satellite technology.

All manuscripts are peer-reviewed to meet the highest standards of scientific literature. Interested authors may submit proposals by email to the series editors or to the relevant Birkhäuser editor listed under "Contacts."

Fernando Sansò • Alberta Albertella

The Probabilistic Vision of the Physical World

A Point of View of Earth Sciences

Birkhäuser

Fernando Sansò
Dipartimento di Ingegneria Civile e Ambientale
Politecnico di Milano
Milano, Italy

Alberta Albertella
Dipartimento di Ingegneria Civile e Ambientale
Politecnico di Milano
Milano, Italy

ISSN 2730-5996 ISSN 2512-3211 (electronic)
Lecture Notes in Geosystems Mathematics and Computing
ISBN 978-3-031-88267-8 ISBN 978-3-031-88268-5 (eBook)
https://doi.org/10.1007/978-3-031-88268-5

The original submitted manuscript has been translated into English. The translation was done using artificial intelligence. A subsequent revision was performed by the author(s) to further refine the work and to ensure that the translation is appropriate concerning content and scientific correctness. It may, however, read stylistically different from a conventional translation.

This book is published under the imprint Birkhäuser, www.birkhauser-science.com by the registered company Springer Nature Switzerland AG
The registered company address is: Gewerbestrasse 11, 6330 Cham, Switzerland

Emphasis is also on a basic element of uncertainty which seems to pervade nature and our knowledge of it. It is expressed by Gödel's incompleteness theorem, by fuzzy logic, by Heisenberg's uncertainty relation, by other random fluctuations and random measuring errors etc, which have fascinated the imagination of mathematicians, physicists, astronomers and geodesists since C.F. Gauss.

H. Moritz in Science, Mind and the Universe [35]

Preface

When I was asked to give a general talk, as suggested by the title of the text, at a conference on "Potentially dangerous natural events: models, uncertainties, communications," sponsored by CNR (Consiglio Nazionale delle Ricerche, Italy) and IUGG (International Union of Geodesy and Geophysics), besides being very pleased, I realized that I had to systematize and put into a single logical framework what I had been teaching for over 30 years at the Politecnico di Milano. Although this was a regular 30-minute talk, I thought that such a vast area required the creation of a written background structure.

The material and ideas that I began to gather grew in my hands and so I thought of involving in the writing of what has become this text a former student of mine and for many years collaborator on the subject, with great discussions on the approach and conclusions, Dr. A. Albertella, co-author of the book.

The subject of the book is typical of the philosophy of science, that is, the relationship between empiricism and scientific knowledge, which we have however tried to contain from the point of view of Earth physics, to avoid excessively widening the scope of the discussion, risking not to reach any clear conclusion.

The theme is therefore the answer to what is the role of experiments in the construction of our vision of the world and how theory and empirical results can be put together, in particular, what validity can be assigned to the predictions in the field of Earth sciences of a theory with a probabilistic basis, which is uncertain.

Therefore, in the first chapter we try to clarify the contribution of Probability Theory to the physical knowledge of the world and its relationship with statistics, a discipline that has long been erroneously seen as isolated from the first. In the second chapter we look for the origins of stochasticity, linked to problems of complexity, instability, and control. In the following chapter we develop the analysis of inference methods for what concerns estimation, which links the observed data to the theoretical model.

In the fourth chapter we consider the problem of statistical inference from the point of view of hypothesis testing, or the falsifiability of the theoretical model based on experimental data. In the following chapter we focus on the relationship between a discrete modeling of phenomena compared to a continuous one, or between models with a finite number of parameters and those with infinite parameters, also called non-parametric. Here the point of view is that between the two approaches, in particular in Earth sciences, there can be no contradiction but one must be seen naturally as the limit of the other.

The sixth chapter is dedicated to the role of machine learning and deep machine learning, as algorithmic support for the broader field of Artificial Intelligence, around which in our opinion a more mystical than scientific vision has developed. In particular, it shows how machine learning is essentially a natural evolution of the probabilistic vision, in which empirical aspects, that is, "learning from data," are emphasized to the extreme. As such, these methods must be considered as useful tools, whose results must however always be verified from the point of view of already consolidated physical knowledge.

The text concludes with four appendices that contain the mathematical recalls necessary for the understanding of the different topics, without interrupting their logical flow.

Consider that the examples presented in the book, although mostly taken from various fields of Earth sciences, are deliberately extremely simple having more an explanatory purpose than a demonstrative one. Also note that the material used for the construction of the text is taken from a wide and well-known literature and therefore is not original, except for the discussion on the Bayesian estimation of a random field (Chap. 5), of the combined estimation from many models (Sect. 4.4) and of Appendix D.

Finally, we want to underline that perhaps it is not a coincidence that a general discussion of this type takes place from the "point of view of Earth sciences." In fact, these themes are in continuity with the historical thread of which we mention, without any claim to completeness, the work of Gauss and Legendre, who developed the first theories of data analysis of astronomical or Earth observations, of H. Jeffreys, with his contribution to the introduction of the Bayesian point of view in Earth sciences [24] and again of A. Tarantola [52], who developed its contents including aspects of Information Theory. Finally, we would like to mention H. Moritz who in the text [35] addresses the same theme of this book, with an approach that finds a solution to the problem of knowledge more in a philosophical vision than in a scientific one.

Milano, Italy Fernando Sansò

Declarations

Competing Interests The authors have no competing interests to declare that are relevant to the content of this manuscript.

Contents

Chapter 1
Probability and Frequency

The chapter, after recalling the axiomatic definition of probability as opposed to that of frequency of events, illustrates their conceptual differences with a number of elementary examples. Subsequently, showing that probabilities and frequencies share some mathematical properties, it is clarified too that zero-frequency and zero-probability cases have a substantially different meaning; this invalidates the von Mises attempt to define probability as limit of frequencies.

1.1 Different Concepts, Same Formulas

If there is a profound change in the history of science in the way of seeing, conceiving, and describing nature, marking a reversal from the deterministic conception, which operates by *abstraction* by establishing formulas that exactly link the quantities at play, this is the stochastic approach (from the Greek $\sigma\tau o\chi\alpha\sigma\tau\iota\kappa\acute{o}\varsigma$ conjectural). In this approach, our knowledge can be described as a cloud of possible systems/values of variables and our interaction with the world consists of a continuous filtering of this cloud, through appropriate grids that assimilate new data, in the constant effort to condense the cloud by forcing it to concentrate around a point in space, which represents the state of the system we are investigating.

The basic structure of this conception is the description of each elementary act of knowledge as a stochastic object, that is, an experiment of which a priori (that is, with the information available up to this moment) we are not able to predict the result; our knowledge is expressed by a family of possible results and a measure of priority among these, that is, a probability.

Formally, a probabilistic model is a representation of the stochastic experiment by means of three elements (see [27]):

F. Sansò, A. Albertella, *The Probabilistic Vision of the Physical World*, Lecture Notes in Geosystems Mathematics and Computing,
https://doi.org/10.1007/978-3-031-88268-5_1

- a set Ω that represents all possible outcomes of the experiment
- a family $\mathcal{A}$ of subsets that are called *events*. A typical realization of the event is expressed by "the result ω of the experiment belongs to the set A of the family $\mathcal{A}$",

$$\omega \in A \tag{1.1}$$

For technical reasons $\mathcal{A}$ must contain $\emptyset$, Ω and

$$\{A_n\} \in \mathcal{A}\,,\ \forall n \Rightarrow \cup_n A_n \in \mathcal{A} \tag{1.2}$$

$$A \in \mathcal{A} \Rightarrow A^c \in \mathcal{A}\,, \tag{1.3}$$

that is, $\mathcal{A}$ is a σ-algebra of subsets of Ω
- a measure on the family $\mathcal{A}$ that is a correspondence $A \to P(A) \in \mathbb{R}$

$$0 \leq P(A) \leq 1 \quad P(\emptyset) = 0 \quad P(\Omega) = 1 \tag{1.4}$$

$$P(\cup_n A_n) = \sum_n P(A_n) \quad (A_n \cap A_m = \emptyset,\ n \neq m) \tag{1.5}$$

$$P(A^c) = 1 - P(A) \tag{1.6}$$

Probability theory is that branch of mathematics that studies the transformations of the probability distribution under any law that constructs a correspondence between $(\Omega, \mathcal{A})$ and any other space.

For example, a random variable (one-dimensional) is a function of ω in $\mathbb{R}$

$$\omega \in \Omega \to x(\omega) \in \mathbb{R} \tag{1.7}$$

that is measurable, that is

$$\forall a\,, \quad \{\omega\,;\ x(\omega) \leq a\} \equiv A \in \mathcal{A}\,; \tag{1.8}$$

this essentially means that we are able to say what is the probability of the event $\{x(\omega) \leq a\}$.

In summary: probability is a measure of the degree of knowledge, or perhaps better of ignorance, of the state of a system and probability theory is the branch of mathematics that studies its transformations.

But where do the properties formalized by Kolmogorov for the definition of probability come from? These have been fixed by transforming into a set of principles the behavior of a conceptually different object and from an intelligent but incorrect definition of probability.

We are talking about the frequencies of a statistical variable. A statistical variable is the cataloging of the elements of a population of N individuals based on one (or more) characteristics, each of which can assume some value of a numerical or qualitative type.

Example 1.1 The individuals in a class have a height (approximated to 5 cm) between 150 and 200 cm. The population has size $N = 50$; the cataloging variable, height H, can assume the 6 values $H_i = 150, 160, 170, 180, 190, 200$. The cataloging consists in counting the number N_i of individuals who are characterized by the value $H = H_i$ $(i = 1 \ldots 6)$. For example:

i	1	2	3	4	5	6
H_i	150	160	170	180	190	200
N_i	2	11	17	18	1	1

The possible values of the classification variable are called argumental values; the numbers N_i are the absolute frequencies.

As mentioned, the statistical variable, that is, the variable used for classification, can also be of a qualitative type.

Example 1.2 On an image of 10^6 pixels we analyze the content of each pixel, to find whether it corresponds to one of the following categories C_i $(i = 1 \ldots 6)$:

C_i	water	bare land	lawn	forest	cultivated land	built land
N_i	$10 \cdot 10^4$	$15 \cdot 10^4$	$20 \cdot 10^4$	$10 \cdot 10^4$	$30 \cdot 10^4$	$15 \cdot 10^4$

It is clear that a number can always be assigned to each quality class, thus relating this example to the previous one. However, it is important to consider that categorical statistical variables have different characteristics from numerical ones; just think about the concept of average. In the second example, what would the "average" of the land use destination correspond to in an image?

However, we observe that it is useful to compare populations of different sizes with each other, as far as the same characteristic is concerned, for example by answering questions like: regarding the distribution of the characteristic X (with argument values $x_1 \ldots x_n$), can the two populations A and B be considered similar?

It is clear that the comparison cannot be based on the absolute numbers $\{N_i\}$ since the two populations can have a different size, so the concept of relative number or frequency has been introduced

$$f_i = \frac{N_i}{N} \,. \tag{1.9}$$

Note that by their nature, frequencies, which are usually expressed in percentages, must satisfy the following relationships:

$$f_i \geq 0 \ \forall i \tag{1.10}$$

$$\sum_{i=1}^{N} f_i = 1 \ (f_i \leq 1) \,. \tag{1.11}$$

Furthermore, consider a subset of the argumental values $\Xi = \{x_1, \ldots x_n\}$, for example $S_{ik} = \{x_i, x_k, i \neq k\}$: it is clear that the frequencies related to S_{ik}, that is the percentage number of individuals for whom $X = x_i$ or $X = x_k$, will be

$$f(S_{ik}) \equiv \frac{N_i + N_k}{N} = f_i + f_k \,. \tag{1.12}$$

The example is immediately generalized to various groupings, that is subsets, of the argumental values.

Given the subset of Ξ

$$S = \{x_i, x_k \ldots x_\ell\} \tag{1.13}$$

it will be

$$f(S) = \frac{N_i + N_k \ldots + N_\ell}{N} = f_i + f_k \ldots + f_\ell \,; \tag{1.14}$$

it follows evidently that given two disjoint subsets S_1 and S_2 we find

$$f(S_1 \cup S_2) = f(S_1) + f(S_2) \,. \tag{1.15}$$

As you can see, if you identify Ω with Ξ and $P(S)$ with $f(S)$, the (1.10), (1.11) and (1.15), combined with the obvious statement that if a subset is made up of no value (the empty set) then the frequency is zero, correspond to the formal properties (1.4) (1.5) and (1.6) of the Kolmogorov definition.

The only difference lies in the fact that, always for a statistical variable of number N, the family of all subsets of Ξ is made up of 2^N elements, that is a finite number, while for a random variable, Ω can have the power of the continuum and a family of subsets that constitute a σ-algebra is generally infinite; the classic example is when $\Omega = \mathbb{R}$ and $\mathcal{A}$ the relative Borel algebra, which is the minimal σ-algebra containing all the open, closed, semi-open intervals of $\mathbb{R}$. Moreover, in addition to the formal identity that clearly shows that the axiomatic properties of probability are borrowed from those of frequency, there is a simple experiment that shows that things cannot go differently.

Take a population of N individuals, each identified by a number from 1 to N and by an argument of the variable X; for example the element with the number j has for X the argument x_j. Naturally, the number of arguments n will be less, or at most equal, to that of the population. Then take a urn containing the numbers from 1 to N and draw "at random" one of these going to see what is the value of X for the corresponding individual. It is clear that, using the old definition of Laplace's probability, that is, that the probability of an event (like that we extract an element with argumental value x_j) is the number of favorable results divided by the number of all possible results, you will have

$$P(X = x_j) = f(X = x_j) , \tag{1.16}$$

that is, probability and frequency numerically coincide, from which then follows the formal identity of the properties.

But do the two entities also coincide from a conceptual point of view? The answer is absolutely NO. A probability is an a priori measure that an event may occur, while a frequency is the percentage of times an event occurs in a population of N individuals, in an absolutely certain way.

A probability is an abstract entity that is defined a priori, a frequency is an entity that derives from a posteriori counting.

Example 1.3 Let's take the most elementary case of a stochastic experiment, the toss of a fair coin, with the two faces traditionally called head and tail. If for example I know that the experiment is conducted in a "correct" way with a perfectly balanced coin, the a priori probability model will be

Ω	head	tail
P	1/2	1/2

Now suppose we toss the coin 10 times independently (Bernoulli sample) obtaining 3 times head and 7 times tail. This multiple experiment has as a representation of the results (that is, of what has happened and can no longer change) the statistical variable

Ξ	head	tail
N	3	7
f	30%	70%

How to compare this result with the previous definition will be the subject of Chap. 4 dedicated to the verification of the model.

There is another conceptually important point that differentiates frequency from probability; the interpretation of a zero frequency event and that of a zero probability event.

If in a statistical variable an argument x_0 has zero frequency, this means that in the count of the elements of the population none are found having the argument $X = x_0$, but in this case x_0 should not even be included in Ξ which collects all the values of X in the population. However, if we move to the probabilities of an event $A \subset \Omega$ stating that $P(A) = 0$ two different conclusions can be found. One case is like the previous one, that is, the event $X \in A$ is impossible and therefore more properly A should be excluded from Ω. A second case is quite different: assuming that Ω has the power of the continuum and that the probability P has a continuous distribution in Ω, that is, for example assuming for simplicity that $\Omega \subset \mathbb{R}^n$, there is a measurable, non-negative function $f(\omega)$ such that

$$P(A) = \int_A f(\omega) d\omega \,, \tag{1.17}$$

then it is clear that if A has zero measure, $m(A) = 0$, then we find also $P(A) = 0$, but this does not mean at all that an extraction ω in A is impossible.

Example 1.4 Consider a square terrace Q, of area 1, under a uniform rain field. Thinking of a probability model for the event

$$\omega = \{ \text{ a drop falls on } A \subset Q \} \,,$$

it is natural to take

$$P(A) = a(A) = \text{ area of } A \,.$$

But then what is the probability of the event

$$A_0 = \{ \text{ the drop falls in the center of } Q \} ?$$

Clearly this event is possible but its probability is zero.

This distinction sweeps away the attempt made by von Mises, in the mid-1900s [57], to give the definition of probability on the basis of the frequency of an event. According to von Mises, the probability of an event A can be defined in the following way. The random experiment whose probability distribution is to be defined is repeated N times independently; the number of times $n(A)$ in which the result falls in A is counted and the relative frequency

$$f_N(A) = \frac{n(A)}{N} \,; \tag{1.18}$$

is defined

$$P(A) = \lim_{N \to +\infty} f_N(A) \,. \tag{1.19}$$

We show that this definition leads to a contradiction. Suppose that the (1.19) is acceptable and suppose that

$$0 < p = P(A) < 1 \,, \tag{1.20}$$

as results from the limit of a sequence of experiments.

First of all, considering as an argumental value the entire infinite sequence of extractions $\{\omega_i, i = 1, 2 \ldots\}$, it is easy to see that this can only have zero probability. In fact, remembering that for a composite event whose components are independent, its probability is the product of the probabilities of the individual events, we have

$$P(\omega_1 \in A, \omega_2 \in A^c, \omega_3 \in A \ldots) = P(\omega_1 \in A) \cdot P(\omega_2 \in A^c) \cdot P(\omega_3 \in A) \ldots = \\ = p(1-p) \cdot p \ldots \tag{1.21}$$

Therefore for the first N extractions that fall k times in A and $N-k$ times outside A

$$P_N = p^k (1-p)^{N-k} ; \tag{1.22}$$

if the (1.20) holds, we also have

$$O < \rho = \max[p, (1-p)] < 1 ; \tag{1.23}$$

the probability P of the (1.21) is

$$P < P_N \tag{1.24}$$

and since $P_N \to 0$ for $N \to \infty$ it must be

$$P = 0 . \tag{1.25}$$

In reality, a stronger statement holds. In fact, based on the mathematical properties of probabilities, a theorem can be proven, called the *strong law of large numbers*, which states that the set of all sequences that admit for $f_N(A)$ the limit p, is a set of probability equal to 1 (see [16]). Therefore a sequence that has for example extractions that all fall in A, that is $f_N(A) = 1, \forall N$, or anyway for which the (1.19) does not hold, can only have zero probability. However, this event *is not impossible*, it just has zero probability. Therefore, if by an unfortunate chance we happened to be in that sequence, we would have had to conclude that $p = 1$, contrary to the initial hypothesis. It is clear that an empirical definition cannot be logically and mathematically consistent; this is the conceptual merit of Kolmogorov who cut this knot by creating the definition of probability in an axiomatic way.

Chapter 2
The Sources of Stochasticity

The chapter examines through a number of simple examples the thread along which an observer is pushed to build a theoretical probabilistic model to describe the physical world. It is found that complexity is always a root of the indeterminacy of the results of experiments; this can manifest itself by the instability of the theoretical model and the impossibility to control the conditions under which the experiment is conducted.

2.1 Complexity, Instability and Control

What are the sources of uncertainty in the result of an experiment, in other words, what is the source of stochasticity? How can it be that a phenomenon apparently governed by deterministic laws gives rise to non-determined results?

The answer is that the above laws, for example the laws of mechanics, apply to models, that is, to *abstract* copies of reality. The process of abstraction consists precisely in the elimination from the object of observation of a series of factors that are considered irrelevant for the construction of an "ideal alias" of the object itself, which exists only in the world of ideas, and it is in this world that the laws apply, or rather that the laws emerge as a factor of regularity in the relationships between the various quantities "that matter" in the experiment.

Let's start with an absolutely elementary example: the reconstruction of the shape of an object. I see a tree and I want to represent its shape in a rational way, that is, expressible with formulas, or on a computer screen:

(a) the subject interacts with the object, namely it sees the tree (see Fig. 2.1)
(b) the subject constructs a rational "abstract" model of the tree (see Fig. 2.2)
(c) the subject verifies the adaptation of the model to the object, and discovers that the two things agree quite well, but that the correspondence between the object

F. Sansò, A. Albertella, *The Probabilistic Vision of the Physical World*, Lecture Notes in Geosystems Mathematics and Computing,
https://doi.org/10.1007/978-3-031-88268-5_2

Fig. 2.1 Interaction between subject and object

Fig. 2.2 Abstraction, creation of the model

Fig. 2.3 Comparison between object and model

and the model is not perfect. There are residuals, that is, details that are in the object and that the model has not captured due to abstraction (see Fig. 2.3).

Since the residuals between reality and model are too complex to be described by a simple model, we equip ourselves for a summary description, that is, stochastic/probabilistic, which identifies characteristics that persist when different "samples" are taken from the set of residuals.

Therefore, a first source of stochasticity seems to be the complexity of the object of knowledge that makes it difficult/impossible to construct an "abstract model" that perfectly corresponds to the object itself. Stochasticity is "rooted" in complexity but is "produced" in the process of knowledge.

Example 2.1 The most famous experiment with a stochastic result seems to contradict the previous statement, as it is apparently very simple, that is, determined by few elements. We are talking about the toss of a coin performed by hand with the coin falling on a table. Few seem to be the elements that determine the mechanics of the result: the speed imparted by the movement of the arm to the center of gravity of the coin, the angular speed imparted by the thumb to the coin, the angle of detachment and the distance between the point of detachment of the coin and the surface of the table. In reality, the speed of the arm is controlled by the force imposed by the muscle, let's say the bicep, which is composed of about 250,000 fibers, each of which can be in tension or not in relation to the nerve impulse sent by the brain. The same can be said for the muscles of the hand that control the thumb and therefore the angular speed of the coin. Finally, the point of detachment and the angle of detachment of the coin, which cannot be *controlled* strictly by the motion of arm and hand during the toss, make the trajectory of the coin before falling on the table more or less long and in that stretch the coin can make one or many turns, finally falling on one or the other face in an unpredictable way.

If we look at the experiment starting from the bottom, the first factor we find as a source of unpredictability of the result is the *instability* of the experiment with respect to the ability to control the conditions in which it is realized. A small error in information about the initial state (detachment of the coin) is enough to obtain one or the other of the two possible outcomes. Going up the chain we see that the error in the initial information is due to a lack of *control* that allows us to say that we are always performing the "same" experiment, when we toss the coin several times, obtaining different results; in turn, the lack of control depends on the complexity of the factors of which we should have knowledge in order to predict the result.

Example 2.2 Wanting to come to an example closer to Earth sciences, we can mention the problem of Deterministic Chaos which finds a turning point, within the Theory of Systems, in the work of Lorenz [30] who studied a simplified form of evolution of the atmosphere, reducing it to a system of only 3 degrees of freedom. The system, in spite of its simplicity, as is known produces a flow of trajectories that, although starting very close, then accumulate towards attractors well distinct from each other; the famous butterfly effect that we report in Fig. 2.4.

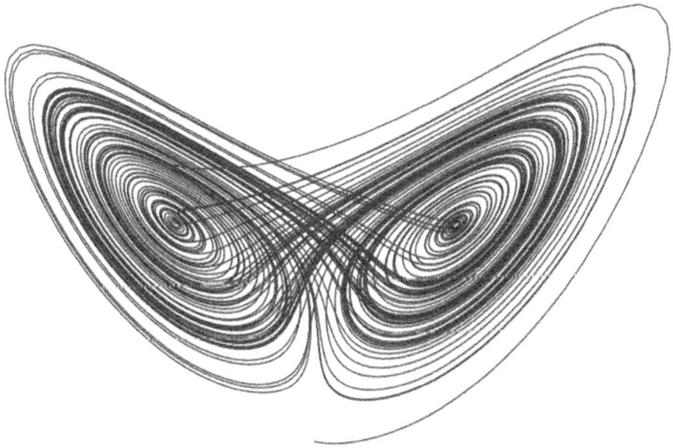

Fig. 2.4 The Lorenz strange attractor

Even more, having a system of a single degree of freedom that follows the evolution equation

$$x_{n+1} = x_n(1 - x_n) \quad 0 \leq x \leq 1 , \tag{2.1}$$

we find it generates a chaotic dynamic behavior that can be described by a probability distribution in phase space.

Does this example seem to go against the idea that complexity is always at the origin of the probabilistic view?

Actually, no. In this case, the complexity originates precisely in the numerical implementation of the system dynamics. If a computer could work with real numbers, there would be no ambiguity in the description of a trajectory; it would be perfectly determined and, if required, time-reversible.

However, "real number" is actually a theoretical concept that always requires a limit process (see also Chap. 5), the complexity of which cannot be controlled by a computer that works with a finite number of digits. It follows that instability, i.e., the divergence of trajectories that have very close initial conditions, always ends up manifesting itself in any numerically feasible solution. Therefore, in this case, as with the coin, the probabilistic view has its roots in complexity and is realized through the instability of the system's evolution.

So it is always complexity that is at the origin of the probabilistic view; this in turn is a tool that serves to express a certain degree of regularity that we find when we examine more closely the part of reality that we have discarded to construct the "model" of the event.

The probabilistic model that we thus construct is actually an idealization of the "sampling" process, or the repetition of the experiment, under the same conditions.

It is important to note that naturally "the same conditions" can only be defined in relation to our control capacity. In other words, we say that a variable $\boldsymbol{x}$ represents a sample of the experiment if each repetition can be exchanged with another, as far as we can understand.

We note that in the process of defining the probabilistic model there can be different approaches, depending on whether one wants to take into account or not a physical content that we can consider governed by a deterministic knowledge of elementary processes that, however, due to their numerosity, are not " controllable", or whether one pushes towards a vision in which the "data itself" dictates a knowledge of the phenomenon, an attitude that is at the basis of the development of modern theory on artificial intelligence, which we will return to.

Two quotes to illustrate this attitude.

From *Lectures on Ergodic Theory*, by P.R. Halmos [17].

> In accordance with classical deterministic mechanics that entire trajectory of the system (the evolution of a system of N particles) can in principle be determined once one instantaneous state is given. In practice we almost never have enough information for such a complete determination. The basic idea of statistical mechanics is to abandon the deterministic study of the state in favor of a statistical (probabilistic) study of an "ensemble" of states.

From *The Elements of Statistical Learning*, by Hastie, Tibshirani, Friedman [19].

> Vast amounts of data are being generated in many fields, and the statistician's job is to make sense of it all: to extract important patterns and trends and understand "what the data says". We call this learning from data.

As seen in the first quote, complexity arises from the numerosity of elementary phenomena ($\sim 6 \cdot 10^{23}$ Avogadro's number) for which, however, we think we know the deterministic law, in the second instead the attitude is to assume to start as a *tabula rasa* as to knowledge of the phenomenon and, through the statistical regularity of repetition, "learn" the laws that govern it.

We will return to this matter in the last chapter, however here it is obligatory, given the generality of the topic, to briefly address a different issue; there has been for over a century a foundation of physics constituted by quantum mechanics.

Quoting from *The Quantum Measurement of Gravity for Geodesists and Geophysicists* [45].

> Let us underline that by necessity the use of probability here must be much more intrinsic to the physical mechanism. [...] Here the probability distribution becomes the road to describe the physical state of the system, even when this is constituted by a single particle.

This statement does not mean that probability is an intrinsic element of objective reality, but that probability is a "tool" (thus a model) with which we construct ideal laws to describe its behavior and predict the result of an experiment conducted on an atomic or subatomic scale.

To quote a more authoritative source, i.e., a founder of quantum theory P.A.M. Dirac in his *The principles of quantum mechanics* [12]:

> The principle of causality (determinism) applies only to a system that is and remains undisturbed. Now if a system is small we cannot observe it without producing a significant disturbance, so we should not expect to find a causal (fixed) relationship between the results of our observation. There will therefore be an inevitable indeterminacy in the prediction of experimental results, the theory being generally capable only of obtaining a certain result when an "observation" is made.

So even in this fundamental area of physics, probability is a representation tool (it lives in the world of ideas), not an intrinsic property of the object.

Chapter 3
Statistical Inference: The Theory of Estimation

The chapter starts building the first leg of the relation between the empirical world of experiments and the theoretical world of models. If we consider a statistical variable as the catalogue of repetitions (sampling) of an experiment described by a probabilistic model (here parametric), we can build a composite model for the sampling, namely a likelihood function. Estimation theory is exactly the tool to go from empirical numbers to estimates of the parameters. Maximum likelihood and Least Squares are analyzed recalling the main properties of the corresponding estimators. Widening the probabilistic model including the parameters as random variables, opens the way to Bayesian Theory, and its approach to estimation, that, more properly, in this context becomes prediction.

3.1 Definition of Inference, Estimator and Likelihood Function

In this and the next section, we deal, albeit briefly, with the existing link between probability and frequency, between casual variable and statistical variable, conceived as an inverse problem: learning from data characteristics of the probability distribution.

The link is constructed through the concept of sample and sample variable. A sample of a random experiment, characterized by its probability distribution, is the set of results of N repetitions, which we here assume to be absolutely independent (Bernoulli sample), of the experiment. If the experiment consists in the observation of a random variable, the sample consists of N numerical values, or of N vectors, when the random variable is multidimensional. As will be clarified in the discussion of Chap. 4, the sample values in themselves tell us nothing; it is only when they are input into the probabilistic (abstract) model that they can give us an indication of

F. Sansò, A. Albertella, *The Probabilistic Vision of the Physical World*, Lecture Notes in Geosystems Mathematics and Computing,
https://doi.org/10.1007/978-3-031-88268-5_3

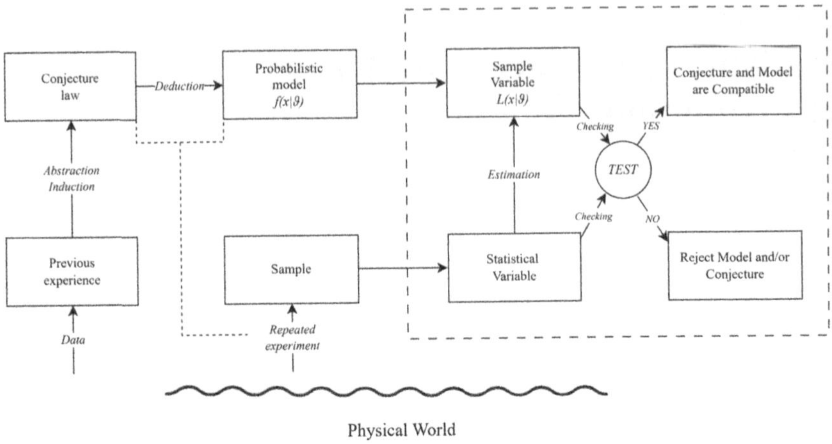

Fig. 3.1 Formation of conjecture, model and statistical inference

the validity of the model itself and of the laws from which it is deductively derived (see Fig. 3.1).

The way in which the sample gives us indications can only be derived from the model itself and is essentially an evaluation of the probability that the observations to which the process of knowledge acquisition leads might not be consistent with the model. Therefore, inference (see Fig. 3.1) is made up of two parts; one is the acquisition of knowledge extracted from the sample to arrive at the complete formulation of the model: this is the theory of estimation. The second is the verification of the result obtained in terms of the probability of its error: this is the theory of statistical tests, or hypothesis testing, of which the so-called decision theory is a branch [7]. In this chapter we deal briefly with the theory of estimation; for this purpose it is necessary that the reader has some basic concepts of probability theory and likelihood. For convenience, these are concisely presented in Appendices A and B. It is advisable for the reader to verify that they know, or at least understand, the topics covered.

To define our problem, let's consider the simplest case of estimation for parametric models. So we have a r.v. (random variable) $\boldsymbol{X}$ in $\mathbb{R}^m$ for which we hypothesize that the likelihood distribution is a certain $L(\boldsymbol{x}|\boldsymbol{\vartheta})$; the difference between $\boldsymbol{x}$ and $\boldsymbol{\vartheta}$ here is that we assume we are able to know a draw $\boldsymbol{x}_0$ from $\boldsymbol{X}$, while we will not have sample values of $\boldsymbol{\vartheta}$. The idea of an estimator $\widehat{\vartheta}(\boldsymbol{X})$ is to construct a r.v. function only of $\boldsymbol{X}$, so that it is known when $\boldsymbol{x}_0$ is known, and that it is plausibly close to ϑ. The value $\widehat{\vartheta}(\boldsymbol{x}_0)$ will then be called an estimate of ϑ.

There are several interesting properties for an estimator $\widehat{\vartheta}(\boldsymbol{X})$ of ϑ; among these we only consider the following:

- **correctness (or unbiasedness)** we say that $\widehat{\vartheta}(\boldsymbol{X})$ is a correct estimator for ϑ if

$$\forall\vartheta\ ,\ E\{\widehat{\vartheta}(\boldsymbol{X})\} = \int L(\boldsymbol{x}|\vartheta)\widehat{\vartheta}(\boldsymbol{x})d_m x \equiv \vartheta\ ; \tag{3.1}$$

this property is realized if the r.v. $\widehat{\vartheta}(\boldsymbol{X})$ is centered at ϑ, for each value of ϑ. More precisely, when the (3.1) holds $\forall\vartheta$ it is said that the estimator is uniformly unbiased,
- **consistency** we say that $\widehat{\vartheta}(\boldsymbol{X})$ is a consistent estimator of ϑ if

$$\lim_{N\to\infty} \widehat{\vartheta}(\boldsymbol{X}) = \vartheta \text{ in probability,} \tag{3.2}$$

i.e. the distribution of $\widehat{\vartheta}(\boldsymbol{X})$ tends to concentrate around the constant value ϑ; different definitions of the limit in (3.2) generate different possible definitions of consistency,
- **efficiency** ideally we would like an estimator that was as "close" as possible to ϑ; by this we mean here that for each ϑ

$$\int [\widehat{\vartheta}(\boldsymbol{x}) - \vartheta]^2 L(\boldsymbol{x}|\vartheta)d_m x = \min \tag{3.3}$$

It is easy to see that in general such an estimator does not exist. We then fall back on the theory of the Minimum Variance Bound (MVB) of Cramer–Rao [7, 8, 44] which establishes the following:

Theorem 3.1 *Consider the score variable (see Appendix B)*

$$U(\boldsymbol{X}|\vartheta) = \frac{L_\vartheta(\boldsymbol{X}|\vartheta)}{L(\boldsymbol{X}|\vartheta)}\ , \tag{3.4}$$

which, as shown in the Appendix, has mean and variance given by

$$E\{U\} = 0\ ,\ \sigma_U^2 = E\{U^2\} \equiv -E\{U_\vartheta\}\ ; \tag{3.5}$$

then any correct estimator $\widehat{\vartheta}(\boldsymbol{X})$ of ϑ has variance such that

$$\sigma_{\widehat{\vartheta}}^2 \geq \frac{1}{\sigma_U^2}\ . \tag{3.6}$$

It is noteworthy that this MVB is reached for sample variables whose likelihood belongs to the exponential family, i.e. have the form

$$L(\boldsymbol{x}|\vartheta) = \exp\{a(\vartheta)\cdot b(\boldsymbol{x}) + c(\boldsymbol{x}) + d(\vartheta)\}\ . \tag{3.7}$$

Many important likelihoods belong to this family, but not all; therefore, an efficiency index of a correct estimator $\widehat{\vartheta}(x)$ has been introduced, as

$$\eta = [\sigma^2_{\widehat{U}} \cdot \sigma^2_{\widehat{\vartheta}}]^{-1} ; \tag{3.8}$$

It is obvious that, for the (3.7),

$$0 \leq \eta \leq 1 \tag{3.9}$$

and that, the closer η is to 1, the higher the effectiveness of the estimator itself.

In this situation, the estimation theory relies on the following approach:

- (a) an optimal estimation "principle" is established for which a solution $\widehat{\vartheta}(X)$ is certain to exist;
- (b) the properties of $\widehat{\vartheta}(X)$ are studied a posteriori.

Here we examine three main estimation criteria.

3.2 Maximum Likelihood Estimators

The principle of maximum likelihood (ML) is probably the most widely used by statisticians who apply it to problems of physics, economics, biology, etc. The method, which originated in the work of R. Fisher [14] in 1922, therefore has an immense bibliographic extension. A brief summary of the asymptotic properties of the related estimator $\widehat{\boldsymbol{\vartheta}}_{ML}(\boldsymbol{X})$ is reported in Appendix B. For the sake of generality we consider here models with a vector parameter $\boldsymbol{\vartheta}$.

The principle is stated as follows: let $\widehat{\boldsymbol{\vartheta}}(\boldsymbol{x})$ be the function derived from the maximum principle

$$\widehat{\boldsymbol{\vartheta}}(\boldsymbol{x}) = \arg\max_{\boldsymbol{\vartheta}} L(\boldsymbol{x}|\boldsymbol{\vartheta}) , \tag{3.10}$$

then

$$\widehat{\boldsymbol{\vartheta}}_{ML}(\boldsymbol{X}) = \widehat{\boldsymbol{\vartheta}}(\boldsymbol{X}) . \tag{3.11}$$

The most interesting properties of $\widehat{\boldsymbol{\vartheta}}_{ML}$ concern the asymptotic behavior, i.e., when the sample size N tends to infinity.

In this case, under not too restrictive conditions, we have that

$$\begin{cases} \widehat{\boldsymbol{\vartheta}}_{ML}(\boldsymbol{X}) \sim \mathcal{N}(\boldsymbol{\vartheta}, C_{\widehat{\boldsymbol{\vartheta}}_{ML}}) \\ C_{\widehat{\boldsymbol{\vartheta}}_{ML}} = C_{\boldsymbol{U}}^{-1} , \end{cases} \tag{3.12}$$

where C_U is the covariance of the vector $\boldsymbol{U}$ defined, in analogy to the one-dimensional case recalled in (3.4), by

$$\boldsymbol{U} = \partial_{\boldsymbol{\vartheta}} \log L\,(\boldsymbol{X}|\boldsymbol{\vartheta})|_{\boldsymbol{\vartheta}=\widehat{\boldsymbol{\vartheta}}_{ML}(\boldsymbol{X})} \,. \tag{3.13}$$

Naturally it is

$$E\{\boldsymbol{U}\} = 0\,,\; C_U \equiv E\{\boldsymbol{U}\boldsymbol{U}^\top\}\,. \tag{3.14}$$

A heuristic demonstration is given in Appendix B for the one-dimensional case only; a more rigorous demonstration can be found for instance in [7, 8].

Therefore, asymptotically $\widehat{\boldsymbol{\vartheta}}_{ML}$ is an unbiased and efficient estimator; but this does not generally happen for a fixed N. Example B.3 shows that in a normal Bernoulli sample, $\widehat{\sigma}^2_{ML}$ is biased. Another advantage of the ML principle is the possibility of treating continuous and discrete variables in the same way, both with respect to $\boldsymbol{x}$ and $\boldsymbol{\vartheta}$.

Example 3.1 Let $\boldsymbol{X}$ be the sample variable whose components, independent, have Bernoulli distributions on the set {0, 1}

$$X_i \sim \begin{cases} 0 & 1-p \\ 1 & p \end{cases}\;; \tag{3.15}$$

it will be

$$P(X_i = x_i \ldots X_m = x_m|p) = p^{x_1}(1-p)^{1-x_1}p^{x_2}(1-p)^{1-x_2}\ldots \tag{3.16}$$

that is, with $x_i = 1$ if the i-th extraction is equal to 1 or equal to 0 in the other case,

$$P(X_1 = x_1 \ldots X_m = x_m|p) = p^{\sum_{i=1}^N x_i}(1-p)^{\sum_{i=1}^N (1-x_i)}\,. \tag{3.17}$$

Let's put

$$n_1 = \sum_{i=1}^N x_i \quad n_0 = \sum_{i=1}^N (1-x_i) = N - n_1 \tag{3.18}$$

and consider the (3.17) as the likelihood related to the continuous parameter p. The ML principle therefore imposes

$$\partial_p \log P = \partial_p(n_1 \log p + n_0 \log(1-p)) = \frac{n_1}{p} - \frac{n_0}{1-p} = 0 \tag{3.19}$$

which gives

$$\widehat{p}_{ML} = \frac{n_1}{n_1 + n_0} = \frac{n_1}{N} , \tag{3.20}$$

a certainly reasonable result. Since n_1 is a sample value from the r.v. $N_1 = \sum_{i=1}^{N} X_i$ and since

$$E\{X_i\} = p \,, \; \sigma^2(X_i) = p(1-p) \tag{3.21}$$

it clearly results

$$E\{\widehat{p}_{ML}\} = \frac{1}{N} E\{N_1\} = p \,, \tag{3.22}$$

that is, in this case $\widehat{p}_{ML}$ is unbiased, and

$$\sigma^2(\widehat{p}_{ML}) = \frac{\sigma^2(N_1)}{N^2} = \frac{Np(1-p)}{N^2} = \frac{p(1-p)}{N} \tag{3.23}$$

which tends to zero, as per the maximum likelihood theory; that is, $\widehat{p}_{ML}$ is, as it should be, a consistent estimator of p.

Example 3.2 Suppose we have a stochastic system composed of two normal populations $\mathcal{N}(\mu_0, \sigma_0^2), \mathcal{N}(\mu_1, \sigma_1^2)$ from one of which a sample $\boldsymbol{x}_0$ of size N is drawn, without knowing whether the tag T of the sample is 0 or 1. The probabilistic model is given by the likelihood

$$L(\boldsymbol{x}|T) = \frac{1}{(2\pi)^{N/2}} \frac{1}{\sigma_T^N} e^{-\frac{1}{2\sigma_T^2} \sum_{k=1}^{N} (x_k - \mu_T)^2} \quad ; \tag{3.24}$$

T is the discrete parameter to be estimated, that is, we want to know whether the population from which x_0 is drawn is $\mathcal{N}(\mu_0, \sigma_0^2)$ or $\mathcal{N}(\mu_1, \sigma_1^2)$. In this case, to create the estimator $\widehat{T}_{ML}$, we need to compare

$$\begin{cases} L_0(\boldsymbol{x}) = \frac{1}{(2\pi)^{N/2}} \frac{1}{\sigma_0^N} e^{-\frac{1}{2\sigma_0^2} \sum_{k=1}^{N} (x_k - \mu_0)^2} \\ \\ L_1(\boldsymbol{x}) = \frac{1}{(2\pi)^{N/2}} \frac{1}{\sigma_1^N} e^{-\frac{1}{2\sigma_1^2} \sum_{k=1}^{N} (x_k - \mu_1)^2} \end{cases} \tag{3.25}$$

at the value $x = x_0$; the ML principle tells us that

$$\begin{cases} \dfrac{L_0(x_0)}{L_1(x_0)} > 1 \Rightarrow \widehat{T}_{ML} = 0 \\ \\ \dfrac{L_1(x_0)}{L_0(x_0)} > 1 \Rightarrow \widehat{T}_{ML} = 1 \, , \end{cases} \tag{3.26}$$

The method thus defined is that of the likelihood ratio which finds application in decision theory and classification. The example will be resumed in the following chapters.

3.3 The Principle of Least Squares

The principle of Least Squares (LS), [44], is originally a purely deterministic principle, which can be stated in full generality as follows:

Let X be an observable r.v. in $\mathbb{R}^m$ and suppose we know that, for some kind of physical or mathematical law, the mean of X, $E\{X\} = x$, is obliged to lie on a manifold $\mathcal{M}$, also called *variety of admissible values*, which we suppose to be of dimension $n < m$, that is, it cannot be any point in $\mathbb{R}^m$ but must necessarily lie on $\mathcal{M}$; for simplicity, we suppose that $\mathcal{M}$ is a smooth and closed variety. Now consider a symmetric matrix W strictly positive definite and associate with it a pseudo-Euclidean metric in $\mathbb{R}^m$, setting

$$d(x, y)^2 = |x - y|_W^2 \equiv (x - y)^\top W(x - y) \, . \tag{3.27}$$

Let x_0 be a sample value of X, i.e., a vector of observed values drawn from X, we take as LS estimator of $x = E\{X\}$ the vector

$$\widehat{x}_{LS} = \arg \min_{x \in \mathcal{M}} d(x_0, x) \, , \tag{3.28}$$

or

$$\forall x \in \mathcal{M} \, , \; (x - x_0)^\top W(x - x_0) \geq (\widehat{x}_{LS} - x_0)^\top W(\widehat{x}_{LS} - x_0) \, . \tag{3.29}$$

From a geometric point of view, the LS estimator is an orthogonal projection of x_0 onto $\mathcal{M}$ (Fig. 3.2). Of course, here orthogonality is understood in the sense of the scalar product $< x, y >_W = x^\top W y$, which leads to the distance (3.27).

Under the assumptions made, such an estimator exists $\forall x_0$, but in general it is not unique (think of $\mathcal{M}$ as a sphere and x_0 its center, with $W = I$).

Of course, in order to numerically implement the (3.28) we need an analytical representation of $\mathcal{M}$; the simplest is a parametric representation of the type

$$x = h(\lambda) \, , \; x \in \mathbb{R}^m \, , \; \lambda \in \mathbb{R}^n \, . \tag{3.30}$$

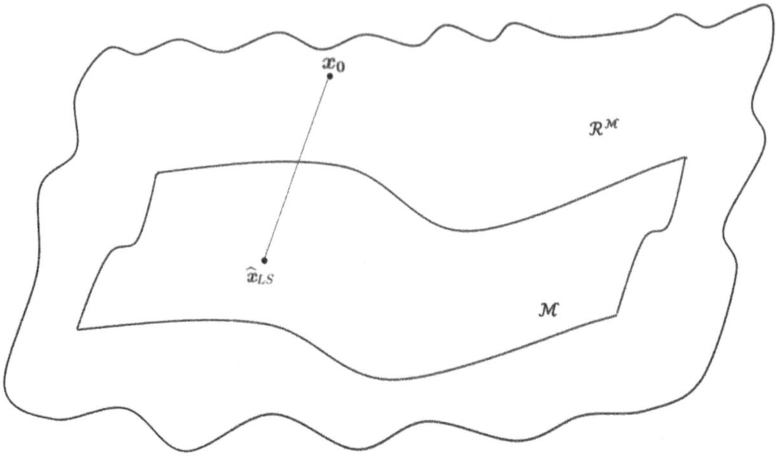

Fig. 3.2 Geometric illustration of the principle of least squares

For simplicity, suppose that the relation

$$\boldsymbol{x} \in \mathcal{M} \Leftrightarrow \boldsymbol{\lambda} \in \mathcal{D} \subset \mathbb{R}^n \tag{3.31}$$

is bijective, i.e., every $\boldsymbol{x} \in \mathcal{M}$ is represented by one and only one $\boldsymbol{\lambda}$, belonging to the domain in $\mathbb{R}^n$; for this reason, the dimension of $\boldsymbol{\lambda}$ must be n, like the dimension of $\mathcal{M}$. Note that in this way to $\boldsymbol{\lambda}$ a role similar to $\boldsymbol{\vartheta}$ of the ML theory is given.

Assuming that $\boldsymbol{h}(\boldsymbol{\lambda})$ is continuous with its first and second derivatives, the necessary condition to find the minimum of the function

$$F(\boldsymbol{\lambda}) = \frac{1}{2} d^2(\boldsymbol{x}_0, \boldsymbol{h}(\boldsymbol{\lambda})) = \frac{1}{2} [\boldsymbol{x}_0 - \boldsymbol{h}(\boldsymbol{\lambda})]^\top W [\boldsymbol{x}_0 - \boldsymbol{h}(\boldsymbol{\lambda})] \tag{3.32}$$

is that

$$\begin{cases} \partial_{\lambda} F = J^\top(\boldsymbol{\lambda}) W [\boldsymbol{x}_0 - \boldsymbol{h}(\boldsymbol{\lambda})] = 0 \\ J(\boldsymbol{\lambda}) = \partial_{\lambda} \boldsymbol{h}(\boldsymbol{\lambda}) , \end{cases} \tag{3.33}$$

where $J(\boldsymbol{\lambda})$ is the Jacobian of $\boldsymbol{h}(\boldsymbol{\lambda})$.

In general, (3.33) admits more than one solution corresponding to multiple possible minima, maxima, or saddle points; to ensure that the solution of (3.33) is a relative minimum, it is necessary to verify that the matrix of the second derivatives of $F(\boldsymbol{\lambda})$ is positive definite

$$\partial_{\lambda} \partial_{\lambda}^\top F(\boldsymbol{\lambda}) > 0 . \tag{3.34}$$

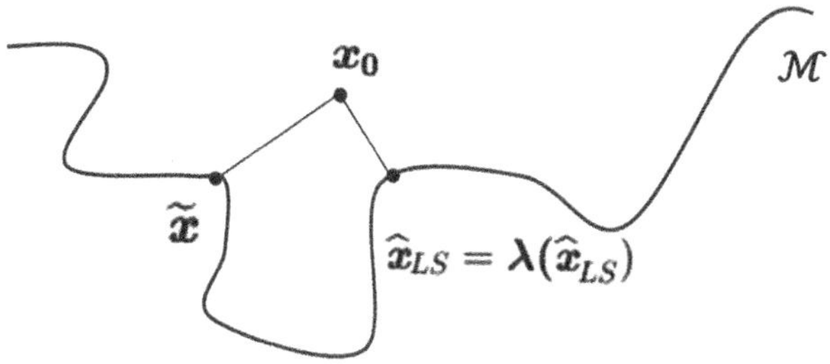

Fig. 3.3 $\widetilde{x}$ relative minimum point of the distance $d(x_0, \mathcal{M})$, $\widehat{x}_{LS}$ absolute minimum point and LS estimator

Even if (3.34) is satisfied, it is still necessary to verify which among the various relative minima gives the absolute minimum (see Fig. 3.3).

Naturally, the estimate $\widehat{\lambda}_{LS}$ then corresponds to the point of $\mathcal{M}$ given by

$$\widehat{x}_{LS} = h(\widehat{\lambda}_{LS}) . \tag{3.35}$$

(3.33) should be changed if $\mathcal{M}$ had a border that should be treated separately; we will not delve into this topic here.

The stochastic properties of $\widehat{x}_{LS}$ and in particular its dispersion should then be studied using the distribution that derives from the orthogonal projection of that of X on $\mathcal{M}$. It is clear that this cannot be done in general terms; however, the widespread use of the LS principle is due to the fact that (3.33) simplifies enormously when $\mathcal{M}$ is a linear variety, i.e.,

$$E\{X\} = h(\lambda) = A\lambda + a , \tag{3.36}$$

with A a constant $m \times n$ matrix and a a constant m vector. In this case, in fact, as is also obvious from the geometric point of view, the function $d(x, \mathcal{M})$ has only one stationary point, corresponding to the minimum of the distance (see Fig. 3.4).

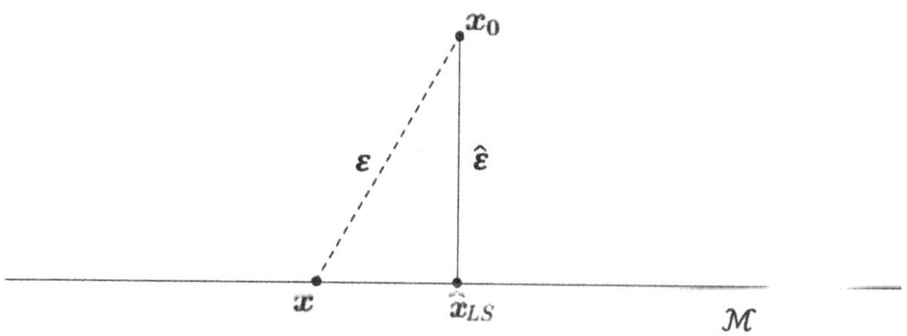

Fig. 3.4 The LS estimator of x_0; x the true mean value of X and ε the discrepancy between the observed value x_0 and the mean x

Note that based on the assumption of bijectivity (3.31), in this case $\mathbb{R}^n \Leftrightarrow \mathcal{M}$, it follows that the matrix A, called *design matrix*, must be of full rank.

Considering that for the linear model (3.36) it results $J(\boldsymbol{\lambda}) = A$, (3.33) becomes

$$A^\top W(\boldsymbol{x}_0 - A\boldsymbol{\lambda} - \boldsymbol{a}) = 0 , \tag{3.37}$$

whose solution is, numerically,

$$\begin{cases} \widehat{\boldsymbol{\lambda}}_{LS} = D_W^{-1} A^\top W(\boldsymbol{x}_0 - \boldsymbol{a}) \\ D_W = A^\top W A . \end{cases} \tag{3.38}$$

The matrix D_W is called *normal matrix* and, from the definition and for the assumptions made, it is strictly positive definite, so that the formula of $\widehat{\boldsymbol{\lambda}}_{LS}$ is admissible.

Naturally, (3.38) provides numerical values of the components of $\widehat{\boldsymbol{\lambda}}_{LS}$ when the observations are $\boldsymbol{x}_0$; the corresponding estimator is

$$\widehat{\boldsymbol{\lambda}}_{LS} = D_W^{-1} A^\top W(\boldsymbol{X} - \boldsymbol{a}) . \tag{3.39}$$

We observe once again that the principle of LS with its solution (3.38) has a purely deterministic character; to each $\boldsymbol{x}_0$ corresponds a

$$\widehat{\boldsymbol{x}}_{LS} = A\widehat{\boldsymbol{\lambda}}_{LS} + \boldsymbol{a} ,$$

i.e., the closest point on the linear variety $\mathcal{M}$, without using in any way the stochastic model of $\boldsymbol{X}$ of which $\boldsymbol{x}_0$ is a draw. It is only when $\widehat{\boldsymbol{\lambda}}_{LS}$ turns into an estimator (see Sect. 3.1), that using the stochastic properties of $\boldsymbol{X}$ one can deduce the properties of $\widehat{\boldsymbol{\lambda}}_{LS}$ as a r.v. .

In particular, it is first of all evident that $\widehat{\boldsymbol{\lambda}}_{LS}$ is a correct estimator of $\boldsymbol{\lambda}$; in fact, due to the linearity of (3.38) and for (3.1)

$$E\{\widehat{\boldsymbol{\lambda}}_{LS}\} = D_W^{-1} A^\top W(A\boldsymbol{\lambda} + \boldsymbol{a} - \boldsymbol{a}) = D_W^{-1}(A^\top W A)\boldsymbol{\lambda} = \boldsymbol{\lambda} . \tag{3.40}$$

Furthermore, for $\widehat{\boldsymbol{\lambda}}_{LS}$ the following classic theorem holds, which resolves the ambiguity introduced in (3.27) by the choice of an arbitrary matrix W.

Theorem 3.2 (Gauss–Markov–Aitken, [1]) *Among all unbiased linear estimators of* $\boldsymbol{\lambda}$, *that is*

$$\widetilde{\boldsymbol{\lambda}} = L\boldsymbol{X} + \boldsymbol{\ell} , \quad E\{\widetilde{\boldsymbol{\lambda}}\} = \boldsymbol{\lambda} \tag{3.41}$$

$\widehat{\boldsymbol{\lambda}}_{LS}$ *is the estimator of minimum variance, if the weight matrix* W *is chosen as*

$$W = C_X^{-1} . \tag{3.42}$$

This means that for any $\widetilde{\boldsymbol{\lambda}}$ of the form (3.41), which is also correct, it necessarily follows that

$$C_{\widetilde{\lambda}} \geq C_{\widehat{\lambda}_{LS}} \,. \tag{3.43}$$

From here on, we will refer to $\widehat{\boldsymbol{\lambda}}_{LS}$ as the least squares estimator when W is chosen as in (3.42). This result, already important in itself, has been significantly strengthened by a recent theorem by Hansen [18] which states that $\widehat{\boldsymbol{\lambda}}_{LS}$ has less dispersed covariance than any estimator $\widetilde{\boldsymbol{\lambda}}$, linear or not, as long as it is unbiased; that is, (3.43) holds whether $\widetilde{\boldsymbol{\lambda}}$ is linear or not, on condition that it is unbiased.

It is worth noting here that if the linear model (3.36) holds, it means that there exists a vector $\boldsymbol{\varepsilon}$ of discrepancy of the observation equations

$$\boldsymbol{X} = A\boldsymbol{\lambda} + \boldsymbol{a} + \boldsymbol{\varepsilon} \,, \tag{3.44}$$

and obviously

$$C_X = C_\varepsilon \,. \tag{3.45}$$

The vector $\boldsymbol{\varepsilon}$ is often considered as a vector of measurement errors, although in many cases this concept must be broadened to include model errors, [52].

In any case, it is often believed that a matrix proportional to C_ε, called the *cofactor matrix* Q, is known for the vector $\boldsymbol{\varepsilon}$

$$C_\varepsilon = \sigma_0^2 Q \,; \tag{3.46}$$

obviously Q must be positive definite and symmetric and σ_0^2 must be a positive constant, which in any case is an unknown of the problem. Therefore, in this case the vector of unknown parameters of the problem is $\boldsymbol{\vartheta} = (\boldsymbol{\lambda}, \sigma_0^2)$, of dimension $p = n + 1$. It is noteworthy that the estimator $\widehat{\boldsymbol{\lambda}}_{LS}$, given by (3.39), with $W = C_X^{-1}$, is actually invariant for the multiplication of C_X by a scalar; it follows that the estimator (3.39) can also be written without knowing σ_0^2, in the form

$$\widehat{\boldsymbol{\lambda}}_{LS} = (A^\top Q^{-1} A)^{-1} A^\top Q^{-1} (\boldsymbol{X} - \boldsymbol{a}) \,. \tag{3.47}$$

Correspondingly, we will have

$$\widehat{\boldsymbol{x}}_{LS} = A\widehat{\boldsymbol{\lambda}}_{LS} + \boldsymbol{a} \,. \tag{3.48}$$

For the propagation of covariance, it is easy to see that the dispersion of $\widehat{\boldsymbol{\lambda}}_{LS}$ and $\widehat{\boldsymbol{x}}_{LS}$ is given by

$$C_{\widehat{\lambda}_{LS}} = \sigma_0^2 (A^\top Q^{-1} A)^{-1} \tag{3.49}$$

and

$$C_{\widehat{x}_{LS}} = \sigma_0^2 A(A^\top Q^{-1} A)^{-1} A^\top . \tag{3.50}$$

To know these matrices, it is therefore necessary to estimate σ_0^2 [44]; it is shown that a correct estimator of σ_0^2, within our model, is given by

$$\begin{cases} \widehat{\sigma}_0^2 = \dfrac{\widehat{\boldsymbol{\varepsilon}}^\top W \widehat{\boldsymbol{\varepsilon}}}{m-n} \\ \\ \widehat{\boldsymbol{\varepsilon}} = \boldsymbol{X} - (A\widehat{\boldsymbol{\lambda}}_{LS} + \boldsymbol{a}) . \end{cases} \tag{3.51}$$

These optimal properties hold for models of linear observation equations, (3.36); but how does this principle behave in front of a nonlinear model like

$$\boldsymbol{X} = \boldsymbol{h}(\boldsymbol{\lambda}) + \boldsymbol{\varepsilon} ? \tag{3.52}$$

To understand that for any $\boldsymbol{h}(\boldsymbol{\lambda})$ the optimal properties of minimum variance stated by the Gauss–Markov–Aitken–Hansen theorem for the linear model are lost, let's take a step back and observe that, also based on Fig. 3.4, it will be

$$|\boldsymbol{x}_0 - \boldsymbol{x}|_W^2 = |\boldsymbol{\varepsilon}|_W^2 = |\widehat{\boldsymbol{\varepsilon}}|_W^2 + |\widehat{\boldsymbol{x}}_{LS} - \boldsymbol{x}|_W^2 \geq |\widehat{\boldsymbol{x}}_{LS} - \boldsymbol{x}|_W^2 , \tag{3.53}$$

That is, $\widehat{\boldsymbol{x}}_{LS}$ is always closer to the mean value $\boldsymbol{x}$ than the observation $\boldsymbol{x}_0$. From here, the mentioned properties of minimum variance descend.

Let's see how this is no longer true in general, in the nonlinear case, with a counterexample.

Example 3.3 Consider the model of observation equations

$$\begin{cases} X_1 = \lambda + \varepsilon_1 \\ X_2 = \lambda^2 + \varepsilon_2 , \end{cases} \tag{3.54}$$

with

$$C_\varepsilon = I ; \tag{3.55}$$

suppose that the theoretical value of $\begin{vmatrix} X_1 \\ X_2 \end{vmatrix}$ is $\boldsymbol{x} = \begin{vmatrix} 1/2 \\ 1/4 \end{vmatrix}$, while the observed vector is

$$\boldsymbol{x}_0 = \begin{vmatrix} -0.5 \\ +0.5 \end{vmatrix} . \tag{3.56}$$

The LS principle in this case reduces to finding the minimum of

$$F(\lambda) = (\lambda + 0.5)^2 + (\lambda^2 - 0.5)^2 \tag{3.57}$$

which has a (unique) solution

$$\lambda_{LS} = -\frac{1}{\sqrt[3]{4}} \cong -0.63 ,$$

that is

$$\begin{cases} x_{1LS} = -0.63 \\ x_{2LS} = +0.40 \end{cases}$$

As you can see

$$|\widehat{\boldsymbol{x}}_{LS} - \boldsymbol{x}| = 1.14 , \quad |\boldsymbol{x}_0 - \boldsymbol{x}| = 1.03 ,$$

that is $\widehat{\boldsymbol{x}}_{LS}$ is further from $\boldsymbol{x}$ than $\boldsymbol{x}_0$.

The phenomenon is illustrated in Fig. 3.5, where the circle, with a radius of 1.18, is also represented, in which 50% of the probability around the peak is collected, in the case that $\boldsymbol{\varepsilon}$ is normally distributed; in other words, $\boldsymbol{x}_0$ was chosen in a high probability area.

Generalizing the example, it is seen that, to avoid a worsening of the estimator $\widehat{\boldsymbol{x}}_{LS}$ of the mean $\boldsymbol{x}$ compared to the observed value, it is necessary that the variety $\mathcal{M}$ has a very small curvature in an area where a high probability (for example 90%, 99%) of having an extraction of the observed vector $\boldsymbol{x}_0$ is concentrated, so that $\mathcal{M}$, seen from $\boldsymbol{x}_0$, is essentially flat in the area where the probability of observations is concentrated. Translated into analytical terms this means that $\boldsymbol{h}(\boldsymbol{\lambda})$ can reasonably be approximated, there where it counts, by a linear variety tangent, at a point $\widetilde{\boldsymbol{x}} =$

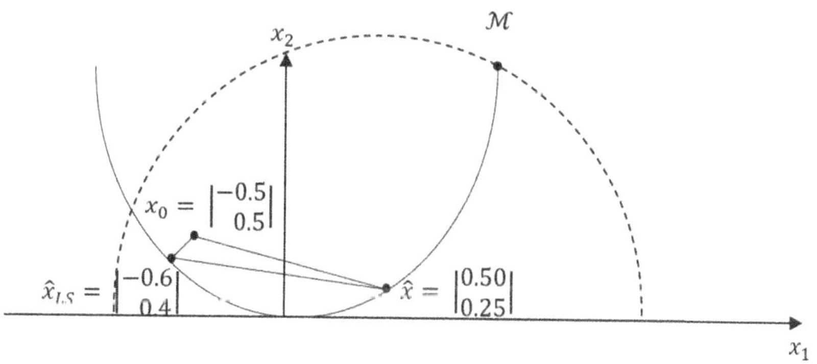

Fig. 3.5 Graphical representation of Example 3.3

$\boldsymbol{h}(\widetilde{\boldsymbol{\lambda}})$, which is known to be an approximation of the solution $\widehat{\boldsymbol{x}}_{LS}$,

$$\boldsymbol{h}(\boldsymbol{\lambda}) \cong \boldsymbol{h}(\widetilde{\boldsymbol{\lambda}}) + J(\widetilde{\boldsymbol{\lambda}})(\boldsymbol{\lambda} - \widetilde{\boldsymbol{\lambda}}) \tag{3.58}$$

to which the already seen linear theory can then be applied. Of course, once the approximated vector $\widehat{\widetilde{\boldsymbol{\lambda}}}_{LS}$ is determined, one can proceed to iterate starting from this as a point of linearization.

We close this brief review of the LS noting that in the case where the vector $\boldsymbol{\varepsilon}$ is normally distributed the ML principle and that of the LS coincide at least as far as the estimate of $\boldsymbol{\lambda}$ is concerned.

In fact, recalling the (A.55) and setting $C_X = \sigma_0^2 Q$, we have

$$\begin{aligned}\log L(\boldsymbol{x}|\boldsymbol{\lambda}, \sigma_0^2) = &-\frac{n}{2}\log 2\pi - \frac{1}{2}\log\det Q - n\log\sigma_0 + \\ &-\frac{1}{2\sigma_0^2}[\boldsymbol{x} - \boldsymbol{h}(\boldsymbol{\lambda})]^\top Q^{-1}[\boldsymbol{x} - \boldsymbol{h}(\boldsymbol{\lambda})]\end{aligned} \tag{3.59}$$

For the ML principle the (3.59) must be maximum, that is by changing sign, $-\log L$ must be minimum, which leads to looking for the minimum

$$\widehat{\lambda}_{LM} = \arg\min_{\boldsymbol{\lambda}} \frac{1}{2}[\boldsymbol{x} - \boldsymbol{h}(\boldsymbol{\lambda})]^\top Q^{-1}[\boldsymbol{x} - \boldsymbol{h}(\boldsymbol{\lambda})] \tag{3.60}$$

$$\widehat{\sigma}_{0ML}^2 = \arg\min_{\sigma_0^2}\{n\log\sigma_0 + \frac{1}{2\sigma_0^2}[\boldsymbol{x} - \boldsymbol{h}(\boldsymbol{\lambda})]^\top Q^{-1}[\boldsymbol{x} - \boldsymbol{h}(\boldsymbol{\lambda})]\}\,. \tag{3.61}$$

As you can see the (3.60) provides the same solution $\widehat{\boldsymbol{\lambda}}$ of the LS principle, that is

$$\widehat{\boldsymbol{\lambda}}_{ML} \equiv \widehat{\boldsymbol{\lambda}}_{LS}\,, \tag{3.62}$$

while the (3.61) in which you can put $\boldsymbol{\lambda} = \widehat{\boldsymbol{\lambda}}_{ML}$, gives

$$\widehat{\sigma}_{0ML}^2 = \frac{[\boldsymbol{x} - \boldsymbol{h}(\widehat{\boldsymbol{\lambda}}_{ML})]^\top Q^{-1}[\boldsymbol{x} - \boldsymbol{h}(\widehat{\boldsymbol{\lambda}}_{ML})]}{n}\,. \tag{3.63}$$

Therefore $\widehat{\sigma}_{0ML}^2$ is biased, as already happened in the estimate of σ^2 for a normal Bernoulli sample; for this reason from the application point of view the (3.51) is always used.

3.4 Bayesian Approach: Minimum Mean Square Error Estimators and Maximum A Posteriori Estimators

A decisive step towards finding an estimator that minimizes the mean square error of estimation, which is not solved within ML theory, can be made by changing the paradigm of the interpretation of observation equations, accepting a Bayesian approach. According to this approach, every variable that enters our knowledge model related to an observation vector $\boldsymbol{x}_0$ must be a r.v. and the state of knowledge related to these variables before making the observations is expressed through their joint probability distribution [6, 9, 19, 24, 51].

So if an experiment involves the variables $\boldsymbol{x}, \boldsymbol{\vartheta}$, of which only the former are observable, there will be a "prior" distribution with density $f(\boldsymbol{x}, \boldsymbol{\vartheta})$ of the r.v. $\boldsymbol{X}, \boldsymbol{\Theta}$, when these are regular continuous variables.

Recalling (A.18) we can always write

$$f(\boldsymbol{x}, \boldsymbol{\vartheta}) = f(\boldsymbol{x}|\boldsymbol{\vartheta}) \cdot f_{\Theta}(\boldsymbol{\vartheta}) \,. \tag{3.64}$$

$f(\boldsymbol{x}|\boldsymbol{\vartheta})$ is the distribution of $\boldsymbol{X}$, assuming we know $\boldsymbol{\vartheta}$, and it is precisely the likelihood function

$$f(\boldsymbol{x}|\boldsymbol{\vartheta}) = L(\boldsymbol{x}|\boldsymbol{\vartheta}) \,, \tag{3.65}$$

which reflects the behavior of measurement errors. For example, for an observation equation model

$$\boldsymbol{X} = \boldsymbol{h}(\boldsymbol{\lambda}) + \boldsymbol{\varepsilon} \tag{3.66}$$

if $f_{\varepsilon}(\boldsymbol{\varepsilon})$ is the distribution of $\boldsymbol{\varepsilon}$, it will be

$$L(\boldsymbol{x}|\boldsymbol{\lambda}) = f_{\varepsilon}[\boldsymbol{x} - \boldsymbol{h}(\boldsymbol{\lambda})] \,. \tag{3.67}$$

The marginal $f_{\Theta}(\boldsymbol{\vartheta})$ describes our prior knowledge of $\boldsymbol{\vartheta}$; usually in Bayesian literature it is indicated as

$$p(\boldsymbol{\vartheta}) = f_{\Theta}(\boldsymbol{x}) \,, \tag{3.68}$$

just to remember that the distribution is "prior" to the knowledge that comes from the observations $\boldsymbol{x}_0$. So we will write (3.64) as

$$f(\boldsymbol{x}, \boldsymbol{\vartheta}) = L(\boldsymbol{x}|\boldsymbol{\vartheta})p(\boldsymbol{\vartheta}) \,. \tag{3.69}$$

From a strictly Bayesian point of view, the solution to the problem of evaluating the introduction of the observation $\boldsymbol{x}_0$ in the prior information expressed by $f(\boldsymbol{x}, \boldsymbol{\vartheta})$

lies in the application of Bayes' theorem represented by (A.33), i.e., the construction of the "posterior" distribution of $\boldsymbol{\Theta}$.

$$f(\boldsymbol{\vartheta}|\boldsymbol{x}_0) = \frac{L(\boldsymbol{x}_0|\boldsymbol{\vartheta})p(\boldsymbol{\vartheta})}{\int L(\boldsymbol{x}_0|\boldsymbol{\eta})p(\boldsymbol{\eta})d\eta} . \tag{3.70}$$

Now it is possible to find methods, in particular Monte Carlo methods, for the generation of the posterior distribution (3.70), or for the generation of samples from this [51], however often one wants to find a representative value for $\boldsymbol{\Theta}$. Given that (see (A.34))

$$\widehat{\boldsymbol{\vartheta}}(\boldsymbol{X}) = E\{\boldsymbol{\Theta}|\boldsymbol{X}\} , \tag{3.71}$$

is a minimum variance (MV) estimator, it may be natural to choose (3.71) as the Bayesian estimate, i.e.,

$$\widehat{\boldsymbol{\vartheta}}_{MV}(\boldsymbol{x}) = \frac{\int \boldsymbol{\vartheta} L(\boldsymbol{x}_0)|\boldsymbol{\vartheta})p(\boldsymbol{\vartheta})d_p\vartheta}{\int L(\boldsymbol{x}_0|\boldsymbol{\vartheta})p(\boldsymbol{\vartheta})d_p\vartheta} . \tag{3.72}$$

Of course, this estimate must be accompanied by a dispersion index which can only be the covariance matrix of the error, conditioned on the observations; that is, given

$$\boldsymbol{\ell}(\boldsymbol{X}, \boldsymbol{\Theta}) = \boldsymbol{\Theta} - \widehat{\boldsymbol{\vartheta}}(\boldsymbol{X}) , \; \boldsymbol{\ell}(\boldsymbol{x}_0, \boldsymbol{\vartheta}_0) = \boldsymbol{\vartheta} - \widehat{\boldsymbol{\vartheta}}_{MV}(\boldsymbol{x}_0) , \tag{3.73}$$

we calculate

$$C_{\widehat{\vartheta}}(\boldsymbol{x}_0) = \frac{\int \boldsymbol{\ell}(\boldsymbol{x}_0, \boldsymbol{\vartheta}_o)\boldsymbol{\ell}(\boldsymbol{x}_0, \boldsymbol{\vartheta})^{\top} L(\boldsymbol{x}_0|\boldsymbol{\vartheta})p(\boldsymbol{\vartheta})d_p\vartheta}{\int L(\boldsymbol{x}_0|\boldsymbol{\vartheta})p(\boldsymbol{\vartheta})d_p\vartheta} . \tag{3.74}$$

A remark is very convenient in the calculation of conditioned averages, like (3.72), (3.74): any multiplicative factor function only of $\boldsymbol{x}_0$ in the $L(\boldsymbol{x}_0|\boldsymbol{\vartheta})$ comes out of the integral at numerator and denominator and therefore can be simplified. The same naturally applies to constants.

The existence of $p(\boldsymbol{\vartheta})$ in the Bayesian formula remains a point of discussion among statisticians. There are three approaches to the determination and use of $p(\boldsymbol{\vartheta})$:

1. $p(\boldsymbol{\vartheta})$ can be derived from a prior stage of knowledge of the variables $\boldsymbol{\vartheta}$; for example, the entries of $\boldsymbol{\vartheta}$ could be coordinates of points previously determined with a certain degree of accuracy and which are now redetermined based on new measurements, $\boldsymbol{x}_0$.
2. Knowing nothing a priori about $\boldsymbol{\vartheta}$, we look for a "non-informative" distribution that therefore does not disturb the estimate $\widehat{\boldsymbol{\vartheta}}$. The argument has been treated by Jeffreys [24] who, based on reasoning about the invariance of estimators,

proposed: for a parameter ϑ, invariant for translation, such as the mean, it is good to use an "improper" uniform distribution over the entire axis, i.e., use

$$p(\vartheta)d\vartheta \equiv d\vartheta \; ; \tag{3.75}$$

for an estimator that must be positive invariant for dilation like the mean square deviation σ, it is good to use an improper distribution by setting

$$p(\vartheta)d\vartheta = \frac{d\sigma}{\sigma} \; . \tag{3.76}$$

3. When instead we have a vague a priori knowledge of $\boldsymbol{\vartheta}$, for example a value $\widetilde{\boldsymbol{\vartheta}}$ that we know is roughly approximate, we can use for $p(\boldsymbol{\vartheta})$ a distribution centered at $\widetilde{\boldsymbol{\vartheta}}$ but very dispersed; for example, we can set, for a p-dimensional vector,

$$p(\boldsymbol{\vartheta}) = \frac{1}{(2\pi)^{p/2}(\widetilde{\sigma})^p} e^{-\frac{1}{2\widetilde{\sigma}^2}|\boldsymbol{\vartheta}-\widetilde{\boldsymbol{\vartheta}}|^2} \tag{3.77}$$

with a suitably high value of $\widetilde{\sigma}$.

We report two elementary examples to understand how to use these recipes.

Example 3.4 We show an application of Jeffreys' recipe 2.

Let $\boldsymbol{X}$ be a Bernoulli sample of size N from a normal distribution with zero mean and variance σ^2, i.e.

$$\boldsymbol{X} \sim \mathcal{N}(0, \sigma^2 I) \; ; \tag{3.78}$$

we set $\sigma^2 = \vartheta$ so that in this case

$$L(\boldsymbol{x}|\vartheta) = \frac{1}{2\pi^{N/2}\vartheta^{N/2}} e^{-\frac{|x|^2}{2\vartheta}} \tag{3.79}$$

We also note that

$$\frac{d\sigma}{\sigma} = \frac{1}{2}\frac{d\sigma^2}{\sigma^2} = \frac{1}{2}\frac{d\vartheta}{\vartheta} \; . \tag{3.80}$$

Applying (3.70), (3.71), (3.72) we can write, simplifying the constants,

$$\widehat{\vartheta}_{MV} = \frac{\displaystyle\int_0^{+\infty} \vartheta \frac{e^{-\frac{|x|^2}{2\vartheta}}}{\vartheta^{N/2}} \frac{1}{2}\frac{d\vartheta}{\vartheta}}{\displaystyle\int_0^{+\infty} \frac{e^{-\frac{|x|^2}{2\vartheta}}}{\vartheta^{N/2}} \frac{1}{2}\frac{d\vartheta}{\vartheta}} \; . \tag{3.81}$$

An integration by parts of the numerator shows that

$$\int_0^{+\infty} \frac{e^{-\frac{|x|^2}{2\vartheta}}}{\vartheta^{N/2}} d\vartheta = \frac{\frac{|x|^2}{2}}{\frac{N}{2}-1} \int_0^{+\infty} \frac{e^{-\frac{|x|^2}{2\vartheta}}}{\vartheta^{N/2+1}} d\vartheta \,, \tag{3.82}$$

so that substituting in (3.81) we find

$$\widehat{\boldsymbol{\vartheta}}_{MV} = \frac{|\boldsymbol{x}|^2}{N-2} \,; \tag{3.83}$$

the result appears reasonable, even though it presents a bias. The estimator is however consistent.

Example 3.5 In this case we will use proposal 3. Suppose $\boldsymbol{X}$ is a Bernoulli sample of size N from a normal distribution with unknown mean and variance 1, i.e.

$$\boldsymbol{X} \sim \mathcal{N}(\mu, I) \,. \tag{3.84}$$

In this case the likelihood is

$$\begin{cases} L(\boldsymbol{x}|\mu) = \frac{1}{(2\pi)^{N/2}} e^{-\frac{1}{2}|\boldsymbol{x}-\mu\boldsymbol{u}|^2} \\ \boldsymbol{u}^\top = [1 \dots 1] \,, |\boldsymbol{u}|^2 = N \,. \end{cases} \tag{3.85}$$

Now if we know a priori that μ does not have a value of order of magnitude greater than 1, we can use a prior

$$p(\mu) = \frac{1}{\sqrt{2\pi}\widetilde{\sigma}} e^{-\frac{\mu^2}{2\widetilde{\sigma}^2}} \,, \tag{3.86}$$

with $\widetilde{\sigma} \gg 1$, or, setting $\eta = \widetilde{\sigma}^{-2}$,

$$p(\mu) = \frac{\sqrt{\eta}}{\sqrt{2\pi}} e^{-\frac{1}{2}\eta\mu^2} \,, \tag{3.87}$$

with $\eta \ll 1$.

Therefore the a priori joint distribution of $(\boldsymbol{X}, \mu)$ will be

$$L(\boldsymbol{x}|\mu) \cdot p(\mu) = \frac{\sqrt{\eta}}{2\pi^{\frac{N+1}{2}}} e^{-\frac{1}{2}[|\boldsymbol{x}-\mu\boldsymbol{u}|^2+\eta\mu^2]} \,. \tag{3.88}$$

Using the identity

$$|\boldsymbol{x}-\mu\boldsymbol{u}|^2+\eta\mu^2=|\boldsymbol{x}|^2-\frac{(\boldsymbol{u}^\top\boldsymbol{x})^2}{N+\eta}+(N+\eta)\left(\mu-\frac{\boldsymbol{u}^\top\boldsymbol{x}}{N+\eta}\right)^2 \tag{3.89}$$

we see that we can set

$$L(\boldsymbol{x}|\mu)p(\mu)=g(\boldsymbol{x})\cdot e^{-\frac{N+\eta}{2}\left(\mu-\frac{\boldsymbol{u}^\top\boldsymbol{x}}{N+\eta}\right)^2}; \tag{3.90}$$

since, as said, the factor $g(\boldsymbol{x})$ simplifies in the construction of the posterior distribution, from (3.90) we directly see that

$$\widehat{\mu}_{MV}\sim\mathcal{N}\left(\frac{\boldsymbol{u}^\top\boldsymbol{x}}{N+\eta},\frac{1}{\sqrt{N+\eta}}\right). \tag{3.91}$$

Therefore

$$\widehat{\mu}_{MV}=\frac{1}{N+\eta}\sum_{i=1}^{N}x_{0i} \tag{3.92}$$

But since $\eta\ll 1$, Eq. (3.92) is a reasonable estimate of the mean. In particular, for $\eta\to 0$, $\widehat{\mu}_{MV}$ tends to the sample mean and therefore to the ML estimate, which also corresponds to a complete prior ignorance of μ.

Now we observe that if $\boldsymbol{\vartheta}$, or a part of its components, is a discrete variable, then $\boldsymbol{\vartheta}_{MV}$ given by Eq. (3.72) is a vector that usually does not fall into one of the points of the lattice of possible values of $\boldsymbol{\vartheta}$, while the posterior distribution (3.70) will assume positive values only for the possible values of $\boldsymbol{\vartheta}$; remember that when $\boldsymbol{\vartheta}$ is discrete the integral in the denominator of Eq. (3.70) will be replaced by a summation and $f(\boldsymbol{\vartheta}|\boldsymbol{x}_0)$ will assume the meaning of a probability rather than that of a probability density.

In this case an estimate, i.e., a representative value of the posterior distribution of $\boldsymbol{\vartheta}$, could advantageously be the one derived from the ML principle, i.e., the maximum a posteriori

$$\boldsymbol{\vartheta}_{MAP}=\arg\max_{\boldsymbol{\vartheta}} f(\boldsymbol{\vartheta}|\boldsymbol{x}_0). \tag{3.93}$$

Another reason why $\boldsymbol{\vartheta}_{MAP}$ is often preferred as an estimator to $\boldsymbol{\vartheta}_{MV}$ is that the calculation of the marginal

$$f_X(\boldsymbol{x})=\int L(\boldsymbol{x}|\boldsymbol{\vartheta})p(\boldsymbol{\vartheta})d_p\vartheta, \tag{3.94}$$

can be very burdensome, if not impossible. This prevents the calculation of $\widehat{\boldsymbol{\vartheta}}_{MV}$ from Eq. (3.72). On the contrary, taking into account that

$$f(\boldsymbol{\vartheta}|\boldsymbol{x}_0) = \frac{L(\boldsymbol{x}_0|\boldsymbol{\vartheta})p(\boldsymbol{\vartheta})}{f_X(\boldsymbol{x}_0)}, \tag{3.95}$$

it can be seen that the search for the maximum with respect to $\boldsymbol{\vartheta}$ of Eq. (3.95) can be carried out even only knowing the numerator, since $f_X(\boldsymbol{x}_0)$ does not depend on $\boldsymbol{\vartheta}$.

Chapter 4
Statistical Inference: Model Verification

In the chapter the construction of the second leg of Statistical Inference is undertaken. Namely, hypotheses testing. After having examinated the epistemic approach to this matter, namely that no testing can "prove" a theory, but only support whether the hypotheses posed are wrong with a certain probability, we attack the subject following the two main approaches in Statistics, the frequentist approach and the Bayesian.

4.1 Experience Can "Falsify" a Model, Never Prove That It Is "Right"

What knowledge does the search for an optimal estimator bring us, as seen in the previous chapter?

Optimization has a behavior similar to that of the great blind worm of the sands of *Dune* (Frank Herbert, [21]), which moves attracted by the noise of those walking in the desert.

Out of metaphor: what does an estimator $\widehat{\vartheta}$, even if optimal, tell us about the stochastic model that we suppose has generated the empirical data from which $\widehat{\vartheta}$ has been derived? The answer is: taken by itself it gives us no indication.

The reason is that $\widehat{\vartheta}$ is indeed derived from an optimization principle, but this in itself does not guarantee that the "optimal" value found is close to the value of ϑ that we were looking for. Optimal does not imply good.

Consider an estimate $\widehat{\vartheta}_{LS}$ of a parameter ϑ based on the least squares principle; by looking at the residuals of the compensation, we can estimate σ_0^2, i.e., the order of magnitude of the errors expected from the model. If the estimate σ_0^2 turns out to be, say, 100 times larger than the value we expect, we begin to think that something is wrong, that there is a discrepancy between empirical reality and model.

F. Sansò, A. Albertella, *The Probabilistic Vision of the Physical World*, Lecture Notes in Geosystems Mathematics and Computing,
https://doi.org/10.1007/978-3-031-88268-5_4

Therefore, it is only by making some prior assumptions about the quantities at play that empirical estimates can provide indications about the goodness of the model.

Of course, an abnormal error of the data with respect to the model can reveal a theoretical error in the formulation of the different stages of the model; for example, we may think that for a part of the data the description of the operation of the measuring apparatus is not at all what was expected (problem of outliers), or we may think that the physical model from which the observation equations are derived is not adequate, not having taken into account some factors that influence the observations. Take as an example an oscillator with nominal frequency f_0 brought into an orbit around the Earth for the measurement of the distance satellite-receiver on Earth (GNSS). Since the oscillator is immersed in a gravitational potential different from that of the receiver on Earth, due to a general relativity effect the frequency seen from Earth is different, by a factor of the order of $5 \cdot 10^{-10}$, causing an error of the order of a centimeter on the distances, clearly visible compared to the phase noise. GNSS systems naturally carry oscillators that automatically correct this effect, [48] §5.6, [53].

In summary, a theory, a mathematical model for the search of an estimate of unknown quantities is never enough to give us knowledge of the phenomenon; only a posteriori verification can tell us if the empirical data "confirm" the theory. Be careful that here "confirm" does not at all mean "prove": since the confirmation can only be probabilistic, it is limited to stating that from the point of view of empirical data the theory is plausible, or that there are no reasons to say that it is wrong, with a high probability.

Of course, in the early decades of the 1900s with the crisis of the pillars of classical physics, mechanics and electromagnetism, caused by the ability to perform increasingly refined and precise measurements, it is clear that this issue of the relationship between knowledge and data could not fail to involve the scientists who were carrying forward that revolution.

So we turn to them to reinforce what has been said above. For example, Albert Einstein, with his ability to synthesize a concept in a sentence, writes in a letter to Max Born in 1926:

> No amount of experiments can prove that I am right, a single experiment can prove that I am wrong.

And again, Max Planck in *The Knowledge of the Physical World: General Characteristics of Physical Laws* in 1943 [38], writes:

> It follows that we will never be able to establish by means of measurements whether a natural law is valid with absolute precision or not. [...] We must therefore concede that, from a logical point of view, the hypothesis that only statistical laws exist in nature is a priori fully justified.

And again, with a more decisive step towards the philosophy that seeks to systematize these concepts in gnoseological terms, we find for example the analysis developed by Karl Popper in *The Logic of Scientific Discovery* of 1934 [40]. Essentially, the philosopher establishes the "principle of falsifiability", stating that a theory, in our case a model, is scientific, hence controllable, only to the extent that it exposes itself to the possibility of being refuted by experiments and observations, applying the logical principle that if A implies B and B is false then A must also be false.

In conclusion, statistical inference, or the formation of knowledge of a phenomenon through a probabilistic model and the acquisition of data, measurements related to the observable quantities of the model itself, stands on two legs: the complete definition of the model, estimating the unknown quantities from those observable, that is the theory of estimation, and the verification of the plausibility of the model itself in the light of observed data, that is the theory of verification, or of statistical tests.

As seen, we have two different conceptions of model formulation: a frequentist one in which the observables $\boldsymbol{X}$ are stochastic because they describe the data acquisition process which is never error-free, while the unknowns $\boldsymbol{\vartheta}$ are constant parameters; the other Bayesian, in which a priori both $\boldsymbol{X}$ and $\boldsymbol{\vartheta}$ are r.v. with their own prior probabilistic knowledge, but, at the time of measurement, $\boldsymbol{X}$ turns into a constant vector $\boldsymbol{x}_0$ of observed values and we have to study the modification of the r.v. $\boldsymbol{\vartheta}$ generated by the knowledge of $\boldsymbol{x}_0$. Correspondingly, we will have two different approaches to the problem of verification, which we will examine in the next two sections.

It can be observed that there is a hot discussion between statisticians of the frequentist school and the Bayesian school in the construction of a better philosophically founded statistical science (see for example [34]). The discussion inevitably revolves around the gnoseological (or epistemic) meaning of probability and the rules for defining Bayesian priors. We do not delve into this discussion here, being open to any theoretical model capable of indicating, in a fully coherent way, the procedure for verifying its own results. In this sense, we look with interest at the effort to unify the two points of view [33], because we are convinced that the more models justify the same result, or similar results, the better the conclusions drawn will be founded.

4.2 Model Verification: Frequentist Approach

What is briefly exposed here is the theory of point tests, of pure significance, and the choice between two hypotheses alternative to each other (see [7]).

The test is a procedure for verifying a hypothesis related to the probabilistic model $L(\boldsymbol{x}|\boldsymbol{\vartheta})$, which is supposed to describe the sample variable $\boldsymbol{X}$ and in which $\boldsymbol{\vartheta}$ is a vector of unknown parameters on which a hypothesis is formulated. The test is constructed through the following steps:

1. We define the hypothesis H_0 which consists of two statements:

$$H_0 \; : \; \begin{cases} L(\boldsymbol{x}|\boldsymbol{\vartheta}) \text{ is the correct model for } \boldsymbol{X} \, , \\ \\ \vartheta = \vartheta_0 \, . \end{cases} \tag{4.1}$$

   Consequently, if H_0 is true, the distribution of $\boldsymbol{X}$ is known, i.e. $L(\boldsymbol{x}|\boldsymbol{\vartheta}_0)$.
2. We create a function $T(\boldsymbol{x})$ of the observables $\boldsymbol{x}$ alone with the following characteristic: calculated $T_0 = T(\boldsymbol{x}_0)$, if $|T_0|$ is a "small" value we consider H_0 reasonable, if $|T_0|$ is large we think that H_0 is wrong.
3. Using $T(\boldsymbol{x})$, we create in Ω two complementary families of sets dependent on a threshold parameter, c,

$$A(c) \equiv \{\boldsymbol{x} \in \Omega \, ; \; |T(\boldsymbol{x})| \le c\} \tag{4.2}$$

$$R(c) = A(c)^c = \Omega \backslash A(c) \tag{4.3}$$

   $A(c)$ is called *acceptance set* of H_0,
   $R(c)$ is called *rejection set* of H_0 or critical zone.
   Note that $A(c)$ is an increasing family with c, while $R(c)$ is decreasing.
4. Since we know the distribution of $\boldsymbol{X}$ according to H_0, $L(\boldsymbol{x}|\boldsymbol{\vartheta}_0)$, if the hypothesis is correct, we can construct, at least in principle, the distribution of the r.v.

$$T = T(\boldsymbol{X}) \, ; \tag{4.4}$$

   therefore for each c, we can determine the probabilities

$$P(|T(\boldsymbol{x})| \ge c) = P(\boldsymbol{X} \in R(c)) = \alpha \, ; \; P(\boldsymbol{X} \in A(c)) = 1 - \alpha \, ; \tag{4.5}$$

   α is called *significance of the test* and, through the (4.5), it is clear that α will be a function of c

$$\alpha = \alpha(c) \, . \tag{4.6}$$

   Since $R(c)$ is decreasing when c increases, so will $\alpha(c)$ and therefore an inverse function can be defined

$$c = c(\alpha) \, , \tag{4.7}$$

   with an appropriate rule when the (4.5) is constant.
5. We set a significance level α; the most commonly used values in statistics are

$$\alpha = 1\% \quad \text{or} \quad \alpha = 5\% \, ; \tag{4.8}$$

   corresponding to α, we will also have defined the two sets

$$A_\alpha = A[c(\alpha)] \; ; \; R_\alpha = R[c(\alpha)] \, . \tag{4.9}$$

6. We conclude the test in the following way: given the fixed value of α,

   – if

$$|T_0| \leq c(\alpha) \equiv \{x_0 \in A_\alpha\} \tag{4.10}$$

   we conclude that H_0 is not refuted by the data, at the significance level α,

   – if

$$|T_0| > c(\alpha) \equiv \{x_0 \in R_\alpha\} \tag{4.11}$$

   we conclude that H_0 is not plausible at the significance level α.

   Recalling the definition (4.5) it is understood that the meaning of α is the probability that H_0 is rejected when instead this hypothesis is true. This eventuality is called a *type I error*.

The test thus outlined is called a two-tailed test, because if $T(\boldsymbol{X})$, as often happens, has a symmetric distribution with a bell shape, then the condition of rejection of H_0, $\{T < -c, T > c\}$, corresponds precisely to two tails that carry an equal probability, that is $\alpha/2$ (see Fig. 4.1).

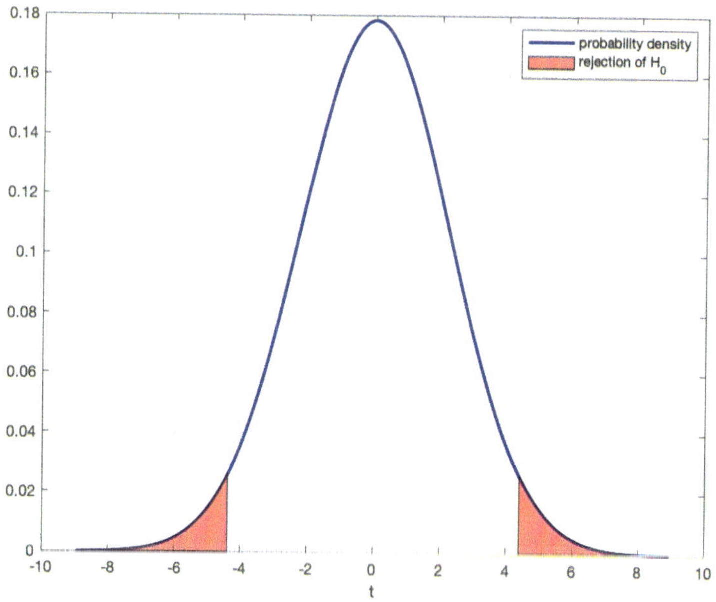

Fig. 4.1 Probability density, symmetric, of $T(\boldsymbol{X})$ and the two tails (where $P = \alpha/2$) that correspond to the rejection of H_0

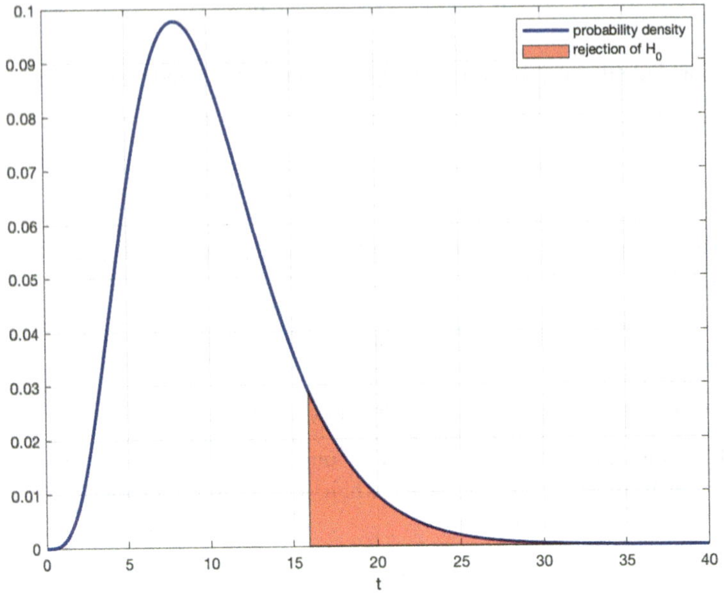

Fig. 4.2 One-tailed test for a positive $T(\boldsymbol{X})$ variable

When $T(\boldsymbol{X})$ is instead strongly asymmetric, for example $T(\boldsymbol{X})$ is structurally positive or null, the area of the t axis that corresponds to the condition of rejection, that is $|T| \leq c$, or $T \leq c$, the test is said to have only one tail that will carry the entire probability α (see Fig. 4.2).

Sometimes for asymmetric distributions it is preferred to define the critical area of the t axis, on which T is distributed, as the area of low probability density, or given a level ℓ it is set

$$P\{f_T(t) < \ell\} = \alpha \tag{4.12}$$

and therefore the acceptance area is that of high probability density (High Probability Density, HPD).

Before moving on to some examples to clarify the formulation of a pure significance test, we observe that the rejection of H_0 (see (4.1)), can occur for one of two reasons: because it is not plausible that $\vartheta = \vartheta_0$, or because it is not plausible that the stochastic model of $\boldsymbol{X}$ is given by $L(\boldsymbol{x}|\boldsymbol{\vartheta})$.

Example 4.1 Let $\boldsymbol{X}$ be a Bernoulli sample drawn from a one-dimensional normal $X \sim \mathcal{N}(\mu, \sigma_X^2)$. Suppose σ_X^2 is known but μ is a parameter to be estimated. Clearly it is

$$\begin{cases} X \sim \mathcal{N}(\mu \boldsymbol{u} \ , \ \sigma_X^2 I) \\ \boldsymbol{u}^\top = [1 \dots 1] \in \mathbb{R}^m \ . \end{cases} \tag{4.13}$$

We want to test the hypothesis

$$H_0 : \ \ \mu = \mu_0 \ . \tag{4.14}$$

Since

$$\widehat{\mu} = \frac{1}{m}\boldsymbol{u}^\top X = \frac{1}{m}\sum_{i=1}^{m} X_i \ , \tag{4.15}$$

or, the sample mean, is a correct estimator of μ, it may make sense to use as a statistical variable to test H_0

$$T(X) = \widehat{\mu} - \mu_0 \ ; \tag{4.16}$$

in fact if the estimate $\widehat{\mu}$ is close to μ_0 we will be induced to consider H_0 true.

Since, as known,

$$\widehat{\mu} \sim \mathcal{N}\left(\mu, \frac{\sigma_X^2}{m}\right) \ , \tag{4.17}$$

if H_0 is true, the distribution of $\widehat{\mu}$ is known, and in particular

$$\widehat{\mu} - \mu_0 \sim \mathcal{N}\left(0, \frac{\sigma_X^2}{m}\right) \sim \frac{\sigma_X}{\sqrt{m}}\mathcal{Z} \ , \tag{4.18}$$

where $\mathcal{Z}$ is the standardized normal of dimension 1. Therefore, given α it will be

$$P\{|\mathcal{Z}| > \mathcal{Z}_{\alpha/2}\} = \alpha \tag{4.19}$$

with $\mathcal{Z}_{\alpha/2}$ a known and tabulated value. Subsequently, the critical zone (rejection of H_0) will be

$$|\widehat{\mu} - \mu| > \frac{\sigma_X}{\sqrt{m}}\mathcal{Z}_{\alpha/2} \ . \tag{4.20}$$

If Eq. (4.20) is verified, we reject H_0, otherwise we accept H_0 with a significance level of α.

Example 4.2 In the linear stochastic model

$$X = A\boldsymbol{\lambda} + \boldsymbol{a} + \boldsymbol{\varepsilon} \tag{4.21}$$

$$E\{\boldsymbol{\varepsilon}\} = 0 \qquad C_\varepsilon = \sigma^2 I \tag{4.22}$$

for which $A, \boldsymbol{a}$ are known constant quantities, while $\boldsymbol{\vartheta} = (\boldsymbol{\lambda}, \sigma^2)$ are parameters, we want to test the hypothesis

$$H_0: \text{ the model (4.21) is correct and } \sigma^2 = \sigma_0^2 \,. \tag{4.23}$$

In particular, the hypothesis that the mean model of $\boldsymbol{X}$

$$E\{\boldsymbol{X}\} = A\boldsymbol{\lambda} + \boldsymbol{a} \tag{4.24}$$

is correct can be rejected because the entire vector $\boldsymbol{a}$ is wrong or because some of its components are grossly incorrect; in this case, it is said that the corresponding measurements contain outliers. Therefore, the test can be the beginning of the search for outliers in the measurements, a practical problem in all disciplines that have to deal with large amounts of data and which has seen significant development in the literature starting from the work of Baarda [2] (see also [7, 44]). We recall that by the principle of least squares, considering that $\boldsymbol{X}$ is normal, if H_0 is true, we have

$$\widehat{\boldsymbol{\lambda}}_{LS} = (A^\top A)^{-1} A^\top (\boldsymbol{X} - \boldsymbol{a}) \sim \mathcal{N}(\boldsymbol{\lambda}, \sigma_0^2 (A^\top A)^{-1}) \tag{4.25}$$

$$\widehat{\varepsilon} = \boldsymbol{X} - (A\widehat{\boldsymbol{\lambda}}_{LS} + \boldsymbol{a}) \tag{4.26}$$

$$\widehat{\sigma}_0^2 = \frac{\widehat{\varepsilon}^\top \widehat{\varepsilon}}{m-n} \sim \frac{\sigma^2}{m-n} \chi^2_{m-n} \,, \tag{4.27}$$

where the χ^2_{m-n} distribution, with $m - n$ degrees of freedom, is well known and tabulated.

Therefore, if H_0 is true, Eq. (4.27) holds with $\sigma^2 = \sigma_0^2$, and since the χ^2 distribution is similar to that of Fig. 4.2, a one-tailed test will be performed, identifying the value $\chi^2_{m-n,\alpha}$ and evaluating the empirical value of $\widehat{\sigma}^2$, for which

$$\widehat{\sigma}^2 > \frac{\sigma_0^2}{m-n} \chi^2_{m-n,\alpha} \quad \text{reject } H_0 \,,$$

$$\widehat{\sigma}^2 < \frac{\sigma_0^2}{m-n} \chi^2_{m-n,\alpha} \quad \text{accept } H_0 \,.$$

As you can see, in designing a test there is a considerable margin in the choice of the statistic $T(\boldsymbol{X})$ and the criteria with which to choose the acceptance and rejection sets. Therefore, one might wonder if it is not possible to optimize the design of the test. This cannot be done based on the concept of testing a hypothesis H_0 alone, but becomes feasible when it is possible to formulate an alternative hypothesis, $H_A : \vartheta = \vartheta_A$. In this case, in fact, a *type II error* comes into play, namely

$$P\,(\text{accept } H_0 | H_A \text{ is true}) = \beta\,. \tag{4.28}$$

The probability $(1 - \beta)$ is called the *power of the test* with respect to the alternative hypothesis $\vartheta = \vartheta_A$ and naturally, one will try to make the power maximum, or β minimum (Fig. 4.3). In terms of the families A_α, R_α already defined in (4.9), we can say that

$$1 - \beta = P\{X \in R_\alpha | \vartheta = \vartheta_A\}\,, \tag{4.29}$$

and this is the quantity to be maximized. In this regard, the fundamental Neyman–Pearson Lemma [7] applies.

Lemma 4.1 *The optimal rejection zone R_α is given by*

$$R_\alpha \equiv \{x\,;\ \frac{L(x|\vartheta_A)}{L(x|\vartheta_0)} > c(\alpha)\}\,, \tag{4.30}$$

$$P\{x \in R_\alpha | H_0\} = \alpha\,.$$

In other words, the zone of maximum power for H_A is the zone of probability α, if H_0 is true, in which it is verified that the likelihood ratio satisfies the inequality

$$\frac{L(x|\vartheta_A)}{L(x|\vartheta_0)} > c(\alpha)\,, \tag{4.31}$$

where $c(\alpha)$ is fixed by the relation

$$P\left\{\frac{L(x|\vartheta_A)}{L(x|\vartheta_0)} > c(\alpha) \mid H_0\right\} = \alpha\,. \tag{4.32}$$

Here after we present an example of how verification theory can be used to answer a question of great importance for Earth: can we actually affirm that there is a process of global warming? The affirmative answer to this question provides fundamental support to the theory of climate change [50].

Let's take for example the data set of the planet's average temperatures from 1851 to 2023 [36, 41]. Upon simple inspection, a clear surge in temperature trends around the 1960–1970 era is distinguishable, corresponding to a tumultuous industrial development and a consequent massive increase in CO_2 emissions into the atmosphere. For simplicity, instead of the original data set, we prefer to switch to a data set of first differences in temperatures, so that the change in slope in the trend of Fig. 4.4 now manifests as a jump in the average (see Fig. 4.5).

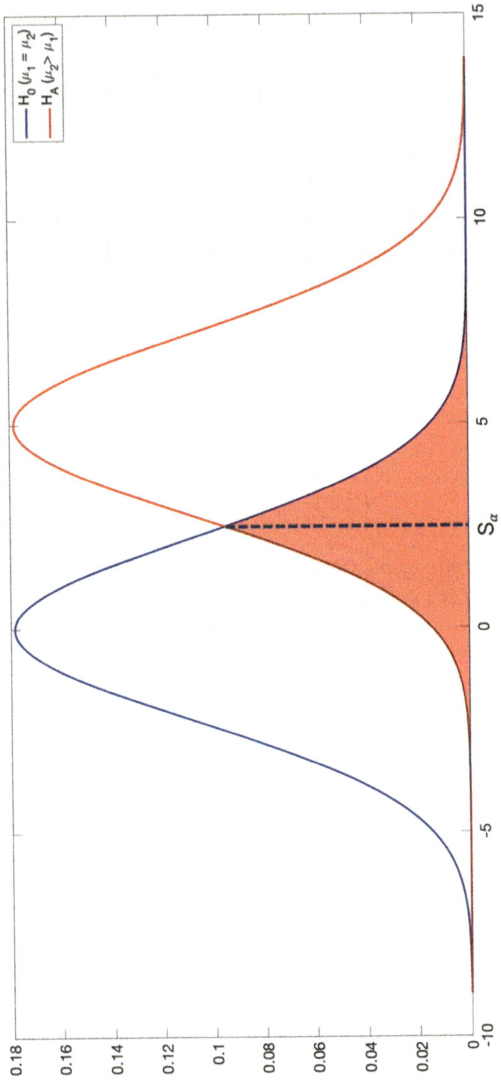

Fig. 4.3 First type errors α, second type β for $\mathcal{M}_2 - \mathcal{M}_1$ and threshold s_α that maximizes $1 - \beta$

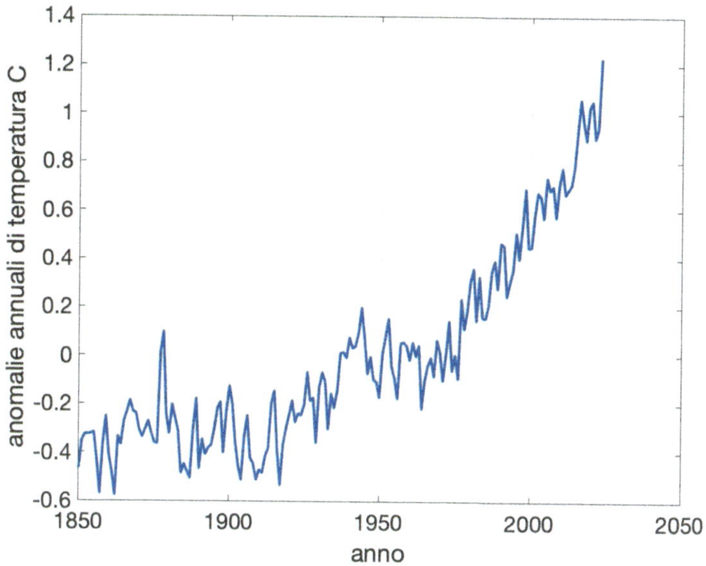

Fig. 4.4 Time series of Earth's average temperatures, from 1851 to 2023

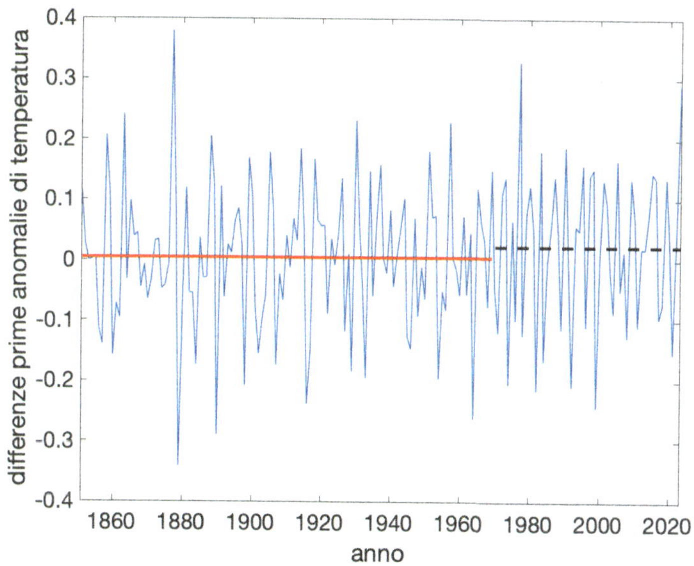

Fig. 4.5 Time series of first differences in Earth's average temperatures, from 1851 to 2023. The horizontal solid line represents the average of the first differences from 1851 to 1969, the horizontal dotted line represents the average from 1970 to 2023

Example 4.3 Considering the data set of annual temperature variations $\{X_i\}$ in Fig. 4.5, we formulate two alternative hypotheses: given

$$\begin{cases} \mu_1 = E\{X_i\}\ i = 1, \ldots, \tau \\ \mu_2 = E\{X_i\}\ i = \tau + 1, \ldots, T \end{cases}, \tag{4.33}$$

assuming that the X_i are independent of each other, we pose

$$\begin{aligned} H_0 &: \mu_1 = \mu_2 \text{ or } \mu_2 - \mu_1 = 0 \\ H_A &: \mu_1 < \mu_2 \text{ or } \mu_2 - \mu_1 > 0 . \end{aligned} \tag{4.34}$$

Given the nature of the test, the natural variable to evaluate the alternative (4.33), (4.34) is clearly

$$T(X) = \mathcal{M}_2 - \mathcal{M}_1 = \frac{1}{N_2} \sum_{i=\tau+1}^{T} X_i - \frac{1}{N_1} \sum_{i=1}^{\tau} X_i . \tag{4.35}$$

We finally assume that the stochastic models of the X_i in the two periods are

$$\begin{cases} X_i = \mu_1 + \varepsilon_{1i}\ i = 1, \ldots, \tau \\ X_i = \mu_2 + \varepsilon_{2i}\ i = \tau + 1, \ldots, T \end{cases} \tag{4.36}$$

with $\varepsilon_{1i}, \varepsilon_{2i}$ normally distributed errors

$$\varepsilon_{1i} \sim \mathcal{N}[0, \sigma_1^2] , \ \varepsilon_{2i} \sim \mathcal{N}[0, \sigma_2^2] . \tag{4.37}$$

In this case, since the data from the two periods are independent of each other, as is known [44], we have

$$\mathcal{M}_2 - \mathcal{M}_1 \sim \mathcal{N}\left[\mu_2 - \mu_1 , \ \frac{\sigma_2^2}{N_2} + \frac{\sigma_1^2}{N_1}\right] . \tag{4.38}$$

Equation (4.38) can also be written as

$$\frac{(\mathcal{M}_2 - \mathcal{M}_1) - (\mu_2 - \mu_1)}{\sqrt{\frac{\sigma_2^2}{N_2} + \frac{\sigma_1^2}{N_1}}} = \mathcal{Z} \tag{4.39}$$

with $\mathcal{Z}$ a standardized normal.

We note that, if we want to maximize the power of the test $(1 - \beta)$, it will be necessary to choose on the real axis, as the rejection zone of H_0, a single tail corresponding to the maximum probability of rejecting H_0 when $H_A : (\mu_2 > \mu_1)$ is true.

Then, if we fix α, the threshold s_α is also fixed.

We note that in (4.39) σ_2^2, σ_1^2 are not exactly known quantities, however, since the two samples are quite large, we can accept the approximation

$$\sigma_2^2 \sim S_2^2 = \frac{1}{N_2}\sum_{\tau+1}^{T}(X_i - \mathcal{M}_2)^2 \,,\ \sigma_1^2 \sim S_1^2 = \frac{1}{N_1}\sum_{1}^{\tau}(X_i - \mathcal{M}_1)^2 \,. \quad (4.40)$$

Therefore, it is possible to determine the threshold s_α of the figure as

$$s_\alpha = \sqrt{\frac{S_2^2}{N_2} + \frac{S_1^2}{N_1}}\,\mathcal{Z}_\alpha \,; \quad (4.41)$$

with $\alpha = 5\%$ in our case it results

$$s_{0.05} = 0.0244 \quad (4.42)$$

Therefore, the test concludes at the significance level of 5% with the statement:

$$\begin{aligned} &\text{if } \mathcal{M}_{20} - \mathcal{M}_{10} > s_{0.05} \quad \text{reject } H_0 \\ &\text{if } \mathcal{M}_{20} - \mathcal{M}_{10} < s_{0.05} \quad \text{accept } H_0 \end{aligned}$$

In any case, the empirical value

$$\mathcal{M}_{20} - \mathcal{M}_{10} = 0.7224$$

is clearly well within the rejection zone of H_0, leading us to the conclusion that the mean μ_2 is plausibly greater than μ_1, i.e., a mechanism of anomalous heating has been triggered, distinct from the random fluctuations that we might consider typical of a natural system. It can be noted that if we use as threshold the observed value $c = 0.7224$, the corresponding "observed" significance is less than $6 \cdot 10^{-50}$.

We observe in conclusion that the alternative hypothesis ($\mu_2 > \mu_1$) is not a simple hypothesis, but a composite one; that is, the value $\mu_2 - \mu_1$ is not fixed by hypothesis, but can take any positive value. Therefore, the role of H_A in this case is rather to fix the optimal zone R_α as a single tail ($\mu_2 - \mu_1 > s_\alpha$).

4.3 Model Verification: Bayesian Approach

This section will be very concise, as it will largely use results already achieved in the previous section, such as the Neyman–Pearson Lemma. We recall that in the Bayesian approach the experimental data $\boldsymbol{x}_0$ is used to generate a posterior distribution of the r.v. $\boldsymbol{\vartheta}$ given by

$$f(\boldsymbol{\vartheta}|\boldsymbol{X} = \boldsymbol{x}_0) = \frac{L(\boldsymbol{x}_0|\boldsymbol{\vartheta})p(\boldsymbol{\vartheta})}{f_X(\boldsymbol{x}_0)} . \tag{4.43}$$

Therefore, it makes no sense to pose the verification of a hypothesis H_0, saying that we want to see if $\boldsymbol{\vartheta} = \boldsymbol{\vartheta}_0$ is rejected by the data or is plausible. The point is that $\boldsymbol{\vartheta}$ is a r.v., while $\boldsymbol{\vartheta}_0$ is a constant. However, we can verify whether $\boldsymbol{\vartheta}_0$ can plausibly be considered a sample drawn from (4.43).

Therefore, we can establish a Bayesian test of pure significance of the hypothesis H_0

$$H_0 \; : \; \boldsymbol{\vartheta}_0 \text{ is a reliable sample of the distribution } f(\boldsymbol{\vartheta}|\boldsymbol{X} = \boldsymbol{x}_0)$$

acting in the following way:

1. we fix a significance level α of the test, i.e., an acceptable probability level of rejecting H_0 when it is true
2. we define in $\mathbb{R}^p$, on which $\boldsymbol{\vartheta}$ is distributed, the two regions

$$A_\alpha \equiv \{\boldsymbol{\vartheta} \in A_\alpha \; , \; H_0 \text{ is accepted}\} \tag{4.44}$$

$$R_\alpha = \{\boldsymbol{\vartheta} \in R_\alpha \; , \; H_0 \text{ is incompatible}\} \tag{4.45}$$

In the absence of further criteria, the acceptance zone A_α can be chosen as the high density (HPD) zone.

We note that when the Bayesian approach is used with an uninformative prior, often the tests of pure significance are similar, if not identical, to the frequentist ones, only the conceptual meaning of $\boldsymbol{\vartheta}_0$ has changed: for the frequentists $\boldsymbol{\vartheta}_0$ is a value imposed a priori on the constant $\boldsymbol{\vartheta}$ to verify its reliability, for the Bayesians $\boldsymbol{\vartheta}_0$ is a potential sample from the posterior of $\boldsymbol{\vartheta}$ to verify if it falls in a predefined significance zone.

Example 4.4 Suppose that $\boldsymbol{X}$ is a Bernoulli sample of size m from a normal distribution with unknown mean and known variance

$$X \sim \mathcal{N}(\mu, \sigma_X^2) . \tag{4.46}$$

Also suppose that the prior of μ is constant, i.e., an improper uninformative distribution. We want to know if, with a given $\boldsymbol{x}_0$ and a fixed α, the value μ_0 can be considered a reliable sample from the posterior of μ. Under the given assumptions, the joint distribution $f(\boldsymbol{X}, \mu)$ is proportional to the likelihood, and the $f(\mu|\boldsymbol{X} = \boldsymbol{x}_0)$ is also proportional to $L(\boldsymbol{x}_0|\mu)$, apart from the marginal $f_X(\boldsymbol{x}_0)$, which however can be considered as a constant with respect to μ. Therefore, it is only a matter of reading the

$$L(\boldsymbol{x}_0|\mu) \equiv \frac{1}{2\pi^{m/2}\sigma_X^m} e^{-\frac{1}{2\sigma_X^2}|\boldsymbol{X}-\mu\boldsymbol{u}|^2}, \tag{4.47}$$

$$(\boldsymbol{u}^\top = [1 \ldots 1])$$

as a distribution in μ.

Using the identity

$$|\boldsymbol{x}_0 - \mu\boldsymbol{u}|^2 = m[(\mu - \widehat{\mu}_0)^2 + \mathcal{S}_0^2] \tag{4.48}$$

$$(\widehat{\mu}_0 = \frac{1}{m}\boldsymbol{u}^\top \boldsymbol{X} = \frac{1}{m}\sum_{i=1}^{m} x_{0i} , \quad \mathcal{S}_0^2 = \frac{1}{m}\sum_{i=1}^{m}(x_{0i} - \widehat{\mu}_0)^2)$$

we can write

$$L(\boldsymbol{x}_0|\mu) = \text{const.}e^{-\frac{m}{2\sigma_X^2}[(\mu-\widehat{\mu}_0]^2+\mathcal{S}_0^2]}; \tag{4.49}$$

Since $\mathcal{S}_0^2$ is a function only of $\boldsymbol{x}_0$ and not of μ, $\exp\left[-\frac{m}{2\sigma_X^2}\mathcal{S}_0^2\right]$ can be included to the multiplicative constant of (4.49) and therefore

$$L(\boldsymbol{x}_0|\mu) = \text{const.}e^{-\frac{m}{2\sigma_X^2}(\mu-\widehat{\mu}_0)^2}. \tag{4.50}$$

If (4.50) must be a probability density in μ, the multiplicative constant is automatically determined by the normalization condition, that is

$$f(\mu|\boldsymbol{X} = \boldsymbol{x}_0) \sim \mathcal{N}\left(\widehat{\mu}_0 , \frac{\sigma_X^2}{m}\right). \tag{4.51}$$

Therefore, given α, the acceptance region of the distribution (4.51) in μ will be

$$|\mu - \widehat{\mu}_0| \leq \frac{\sigma_X}{\sqrt{m}} Z_{\alpha/2} , \tag{4.52}$$

that is H_0 will be accepted if

$$|\mu_0 - \widehat{\mu}_0| \leq \frac{\sigma_X}{\sqrt{m}} Z_{\alpha/2} . \tag{4.53}$$

Recalling (4.18), (4.20), it is seen that the acceptance criterion of H_0 in this case is identical to the frequentist one.

We observe, by giving a final example, that the coincidence of the Bayesian and frequentist criteria is strictly linked to the assumption of a non-informative prior, while if a vague prior information is used, this ends up influencing the Bayesian test.

Example 4.5 Let X be as in Example 4.4, while the prior of μ is

$$p(\mu) \sim \mathcal{N}(\widetilde{\mu}, \widetilde{\sigma}^2) \tag{4.54}$$

with $\widetilde{\sigma}^2$ large so that (4.54) represents a coarse information that μ is centered at $\widetilde{\mu}$ but with a large dispersion.

Therefore in this case

$$f(\boldsymbol{x}|\mu) = \text{cost}\, e^{-\frac{1}{2}\left[\frac{1}{\sigma_X^2}|\boldsymbol{x}-\mu\boldsymbol{u}|^2+\frac{1}{\widetilde{\sigma}^2}(\mu-\widetilde{\mu})^2\right]} \tag{4.55}$$

Now we use the identity (see (4.48))

$$\frac{1}{\sigma_X^2}|\boldsymbol{x}-\mu\boldsymbol{u}|^2+\frac{1}{\widetilde{\sigma}^2}(\mu-\widetilde{\mu})^2 = \frac{m}{\sigma_X^2}[(\mu-\widehat{\mu}_0)^2+S_0^2]+\frac{1}{\widetilde{\sigma}^2}(\mu-\widetilde{\mu})^2 =$$

$$= \frac{1}{\overline{\sigma}^2}\mu^2 - 2\mu\frac{\overline{\mu}}{\overline{\sigma}} + \text{function}(\boldsymbol{x}) \tag{4.56}$$

where

$$\overline{\sigma}^2 = \left(\frac{m}{\sigma_X^2}+\frac{1}{\widetilde{\sigma}^2}\right)^{-1}, \quad \overline{\mu} = \left(\frac{m}{\sigma_X^2}\widehat{\mu}_0+\frac{1}{\widetilde{\sigma}^2}\widetilde{\mu}\right)\overline{\sigma}\,. \tag{4.57}$$

The (4.56), taking into account that functions only of $\boldsymbol{x}$ at exponent can be absorbed by the multiplicative constant, shows that

$$f(\mu|\boldsymbol{X}=\boldsymbol{x}_0) \sim \mathcal{N}(\overline{\mu}, \overline{\sigma}^2)\,. \tag{4.58}$$

As can be seen, $\overline{\mu}$ in this example is a weighted average between $\widehat{\mu}_0$, that is the sample mean that we found in (4.51), and $\widetilde{\mu}$ from the prior information. The weights are respectively $\frac{m}{\sigma_X^2}$ and $\frac{1}{\widetilde{\sigma}^2}$ therefore, if $\widetilde{\sigma}^2$ is large compared to $\frac{\sigma_X^2}{m}$, the solution $\overline{\mu}$ is a perturbation of $\widehat{\mu}_0$. The same happens for $\overline{\sigma}^2$ compared to $\frac{\sigma_X^2}{m}$.

The test for

$$H_0:\ \mu_0 \quad \text{is a sample from } f(\mu|\boldsymbol{X}=\boldsymbol{x}_0)$$

is then done by accepting H_0 if

$$|\mu_0-\overline{\mu}| \le \overline{\sigma} Z_{\alpha/2} \tag{4.59}$$

Finally, we consider the simple alternative between two hypotheses

$$H_0 \; : \; \vartheta = \vartheta_0 \, , \; H_A \; : \; \vartheta = \vartheta_A \, . \tag{4.60}$$

From the Bayesian point of view, one can ask whether ϑ_0 or ϑ_A is more plausible as a sample drawn from $f(\vartheta|X = \boldsymbol{x}_0)$. Also based on the Neyman–Pearson Lemma, it is clear that it will be useful to use the ratio between $f(\vartheta_0|X = \boldsymbol{x}_0)$ and $f(\vartheta_A|X = \boldsymbol{x}_0)$; the larger of the two will be the most plausible. Therefore, the test is done based on the index

$$I(\vartheta_0, \vartheta_A) = \frac{L(\boldsymbol{x}_0|\vartheta_A)p(\vartheta_A)}{L(\boldsymbol{x}_0|\vartheta_0)p(\vartheta_0)} \, , \tag{4.61}$$

resulting in

$$\begin{aligned} I > 1 &\rightarrow \text{ choose } H_A \\ I < 1 &\rightarrow \text{ choose } H_0 \, . \end{aligned}$$

4.4 Statistical Inference Obtained From the Use of Different Models

Often the same observation vector $\boldsymbol{x}$ is considered a sample of an observable X for which different schools construct models, in the context of this book always parametric, obtained from different hypotheses of choice of variables to include in the model and those to neglect or even the choice not to consider in the calculation some terms that are judged irrelevant based on a priori evaluations of their order of magnitude.

Therefore, the same reality, the observations $\boldsymbol{x}$, is tentatively described as sampling from M different likelihoods, or

$$L_m(\boldsymbol{x}|\vartheta_m) \quad m = 1, \ldots, M \, . \tag{4.62}$$

It is good to observe here that the models (4.62) can be different from each other even in the type and number of parameters, which is why the models themselves cannot be simply compared based on the dispersion of the estimate of these. In any case, wanting to build a predictive model, the best possible, for example to be able to make further extractions and therefore simulate a series of future events, we can here assume two different attitudes that also lead to different problems:

(1) we try to understand if among the models there is one better than the others, so to speak the most correct, in representing the data vector $\boldsymbol{x}$,

(2) we assume that all models have their own informative content on the behavior of the variable X and therefore we look for their optimal combination to create the best predictive model.

A Bayesian approach, already seen in Sect. 4.3, provides us with an answer to the two problems.

1. We introduce a vector variable "indicator"

$$W = \{W_m, m = 1, \ldots, M\} \tag{4.63}$$

in which the individual components are binary variables, that is they can only assume the values (0, 1). We then assume that the W_m are linked to each other by the relationship

$$\sum_{m=1}^{M} W_m = 1 \;, \tag{4.64}$$

that is, only one of the W_m can assume the value 1 and the corresponding value indicates the "correct" model. If there is no prior information on the best model, a non-informative distribution can be taken for the vector $\boldsymbol{W}$

$$P_m = P(W_m = 1) = \frac{1}{M} \;. \tag{4.65}$$

With the introduction of $\boldsymbol{W}$ we can define a likelihood

$$L(\boldsymbol{x}|\ldots\boldsymbol{\vartheta}_m\ldots,\boldsymbol{w}) = \sum_{m=1}^{M} W_m L_m(\boldsymbol{x}|\boldsymbol{\vartheta}_m) \tag{4.66}$$

which includes all the models (4.62), choosing that for which $W_m = 1$. Our problem then is to find the posterior distribution of the indicators W_m, that is, of the vector $\boldsymbol{W}$.

With the rule of Bayes we will have

$$P(W_m = 1, \boldsymbol{\vartheta}_m|\boldsymbol{X}) = \frac{L(\boldsymbol{X}|\boldsymbol{\vartheta}_m)P(\boldsymbol{\vartheta}_m, W_m)}{f_X(\boldsymbol{x})} \tag{4.67}$$

and therefore, taking a non-informative prior model

$$P(\boldsymbol{\vartheta}_m, W_m) \div \frac{1}{M} \;, \tag{4.68}$$

we find

$$P(W_m = 1|\boldsymbol{X}) \div \frac{H_m(\boldsymbol{x})\frac{1}{M}}{f_X(\boldsymbol{x})} \;, \tag{4.69}$$

where it is set

$$H_m(\boldsymbol{x}) = \int L(\boldsymbol{x}|\boldsymbol{\vartheta}_m)d\boldsymbol{\vartheta}_m \,. \tag{4.70}$$

If we decide to estimate W_m based on the maximum a posteriori (MAP) principle we can set

$$\widehat{m} = \arg\max_m H_m(\boldsymbol{x}) \,. \tag{4.71}$$

The (4.71) tells us which is the best model for the interpretation of the data $\boldsymbol{x}$. Of course we could reasonably choose the model $\widehat{m}$ and discard the others, that is to say we decide that the model $\widehat{m}$ is the right one, only in the case that the $P(W_{\widehat{m}}|\boldsymbol{x})$ is quite high for example greater than 90%; otherwise we can only say that it is the most likely.

2. In this case we assume a more pragmatic attitude, observing that beyond the simplicity of the symbols in the formulas, in reality each analysis of $\boldsymbol{x}$ with a model $L_m(\boldsymbol{x}|\boldsymbol{\vartheta}_m)$ leads to an estimate $\widehat{\boldsymbol{\vartheta}}_m$ of $\boldsymbol{\vartheta}_m$ as a result of a large work, often also numerical. We therefore leave to the literature the more in-depth discussion of a compendium model of the type

$$L(\boldsymbol{x}|\ldots\boldsymbol{\vartheta}_m\ldots,\boldsymbol{w}) = \sum_{m=1}^{M} W_m L_m(\boldsymbol{x}|\boldsymbol{\vartheta}_m) \tag{4.72}$$

in which W_m can be interpreted as (unknown) prior probabilities of the $m-$th model, (see for example [33]).

Here we will deal with the more practical problem, and also easier, of combining the models in which $\widehat{\boldsymbol{\vartheta}}_m$ have already been determined or the only search for the "weights" $\boldsymbol{W}$ that optimize the combination

$$L(\boldsymbol{x}|\boldsymbol{w}) = \sum_{m=1}^{M} W_m L_m(\boldsymbol{x}|\widehat{\boldsymbol{\vartheta}}_m) \,, \tag{4.73}$$

where the vectors $\widehat{\boldsymbol{\vartheta}}_m$ should be considered as constants. In this case too, we use a Bayesian approach. We note that in (4.73) the weights, in order for $L(\boldsymbol{x}|\boldsymbol{w})$ to be a probability density in $\boldsymbol{x}$, must satisfy the usual convexity condition

$$0 \le W_m \le 1, \tag{4.74}$$

$$\sum_{m=1}^{M} W_m = 1. \tag{4.75}$$

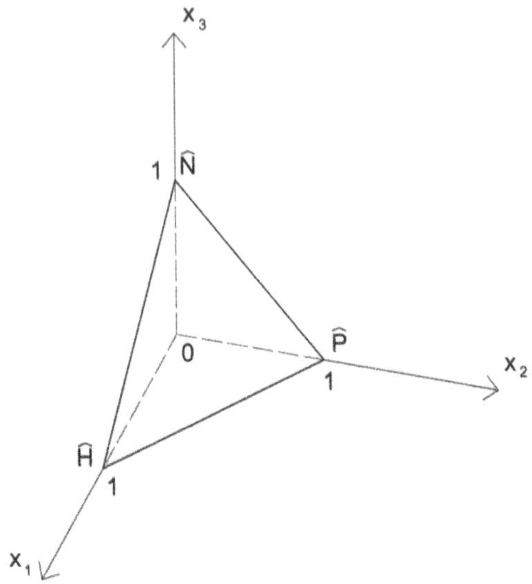

Fig. 4.6 The simplex S_M $(\hat{H}\hat{N}\hat{P})$ for $S_M = 3$, and its base B_M, $(\hat{H}O\hat{P})$, $B_M = S_{M-1}$

That is, the vector $\boldsymbol{W}$ must belong to the simplex S_M which has as vertices the points with coordinates 1 on the axes of $\mathbb{R}^M$ (see Fig. 4.6)

It therefore immediately becomes clear that the non-informative prior distribution $p(\boldsymbol{W})$ must have the form

$$p(\boldsymbol{w}) = \frac{1}{V_M}\chi_M(\boldsymbol{w}) , \tag{4.76}$$

with V_M the $(M-1)$-dimensional volume of S_M and $\chi_M(\boldsymbol{w})$ the indicator of S_M, that is

$$p(\boldsymbol{w}|\boldsymbol{x}) = \begin{cases} 1 & \boldsymbol{w} \in S_M \\ 0 & \boldsymbol{w} \in (\mathbb{R}^M - S_M) . \end{cases} \tag{4.77}$$

So to arrive at the Bayesian estimate of $\boldsymbol{W}$ we just need to write the Bayes formula

$$p(\boldsymbol{w}|\boldsymbol{x}) = \frac{L(\boldsymbol{x}|\boldsymbol{w})p(\boldsymbol{w})}{\int L(\boldsymbol{x}|\boldsymbol{w})p(\boldsymbol{w})d_M\boldsymbol{w}} \tag{4.78}$$

and then calculate the mean of $\boldsymbol{W}$, according to (3.71), with L given by (4.73) and p given by (4.76).

To perform this calculation it is useful to first determine the marginal distributions, $f_m(w_m)$, of (4.76). We immediately notice that for reasons of symmetry these marginals must all be equal, $f(w_m)$. Taking advantage of the fact that a uniform distribution on S_M projects onto a uniform distribution on B_M (see Fig. 4.6), that is, that $(W_1, \ldots, W_{M-1})$ have a constant density on S_{M-1}, and that on S_M it must hold

$$W_M = 1 - \sum_{m=1}^{M-1} W_m \,, \tag{4.79}$$

it is easy to see that the distribution function of W_M is given by

$$0 \le c \le 1, \quad P(W_M \ge c) = 1 - F_{W_M}(c) = P(\sum_{m=1}^{M-1} W_m \le 1 - c) =$$
$$= (1 - c)^{M-1} \,. \tag{4.80}$$

Therefore, for the probability density of W_M, and therefore of all the components W_m, we have

$$f(w_m) = (M - 1)(1 - w_m)^{M-2} \,. \tag{4.81}$$

From here it is trivial to calculate

$$\mu = E\{W_m\} = \frac{1}{M} \,, \tag{4.82}$$

$$q = E\{W_m^2\} = \frac{2}{M(M+1)} \,. \tag{4.83}$$

Note that in (4.82), (4.83) the average operator is always the same because it refers to the same probability density function. It is still useful to calculate the index

$$r = E\{W_i W_k\} \,, \quad (i \neq k) \tag{4.84}$$

also noting that r does not depend on i and k, as long as $i \neq k$. Without the need to calculate the joint marginal of W_i, W_k it is sufficient to remember (4.75) so that we find

$$r = E\{W_i W_k\} = E\{W_i \left(\frac{1}{M-1} \sum_{k \neq i} W_k \right)\} =$$
$$= \frac{1}{M-1} E\{W_i(1 - W_i)\} = \frac{\mu - q}{M-1} = \frac{1}{M(M+1)} \,. \tag{4.85}$$

To proceed, it is convenient to set, for brevity,

$$L_m(\boldsymbol{x}, \widehat{\vartheta}_m) = K_m(\boldsymbol{x}) \tag{4.86}$$

Remembering (4.78), we can now calculate the estimate of $\boldsymbol{W}$, component by component, as

$$\begin{aligned}\widehat{W}_k &= E\{W_k|\boldsymbol{x}\} = \frac{\sum_{m=1}^{M}(\int w_m w_k p(\boldsymbol{w})d\boldsymbol{w})K_m(\boldsymbol{x})}{\sum_{m=1}^{M}(\int w_m p(\boldsymbol{w})d\boldsymbol{w})K_m(\boldsymbol{x})} = \\ &= \frac{r\sum_{m\neq k} K_m(\boldsymbol{x}) + qK_k(\boldsymbol{x})}{\mu\sum_{m=1}^{M} K_m(\boldsymbol{x})} = \\ &= \frac{1}{M+1}\left(1 + \frac{K_k(\boldsymbol{x})}{\sum_{m=1}^{M} K_m(\boldsymbol{x})}\right). \end{aligned} \tag{4.87}$$

As can be seen, the estimates of the W_m result from very simple formulas and it is pleasing to note that the $\widehat{W}_m$ thus estimated satisfy, as they must, the constraints $\widehat{W}_m \geq 0$, $\sum_{m=1}^{M} \widehat{W}_m = 1$.

Of course, the probabilistic model of $\boldsymbol{x}$ that summarizes all the models $m = 1, \ldots, M$ will be given by

$$\boldsymbol{x} \sim \sum_{m=1}^{M} \widehat{W}_m K_m(\boldsymbol{x}) ; \tag{4.88}$$

this model can then be used to simulate new extractions from $\boldsymbol{X}$.

Example 4.6 Let's see an application of the theory explained in the section to a particularly simple case of combination of different models.

Suppose that $\boldsymbol{X}$ has been modeled with M linear normal models, namely

$$L_m(\boldsymbol{x}|\boldsymbol{\lambda}_m) = \mathcal{N}(A_m\boldsymbol{\lambda}_m, \sigma_{0m}^2 I_{Nm}) ; \tag{4.89}$$

let's assume, for simplicity of the example, the σ_{0m}^2 are known and that the design matrices A_m are all of full rank so that the normals

$$D_m = (A_m^\top A_m) \tag{4.90}$$

are invertible. We observe that in general for different m, $\boldsymbol{\lambda}_m$ will have dimension N_m and these are not necessarily equal. As already seen in Example 4.2 the $\boldsymbol{\lambda}_m$ are estimated (identically with the principle of maximum likelihood or least squares) through the formula

$$\widehat{\boldsymbol{\lambda}}_m = D_m^{-1} A_m^\top \boldsymbol{x} . \tag{4.91}$$

As we have seen, problems **(1)** and **(2)** are explicitly solved once the functions $H_m(\boldsymbol{x})$ and $K_m(\boldsymbol{x})$ are known.

For problem **(1)**: we start from the well-known Pythagorean decomposition

$$|\boldsymbol{x} - A_m \boldsymbol{\lambda}_m|^2 = |\boldsymbol{x} - A_m \widehat{\boldsymbol{\lambda}}_m|^2 + |A_m \widehat{\boldsymbol{\lambda}}_m - A_m \boldsymbol{\lambda}_m|^2 . \tag{4.92}$$

We call $\widehat{\boldsymbol{v}}_m$, the vector of estimated residuals,

$$\widehat{\boldsymbol{v}}_m = \boldsymbol{x} - A_m \widehat{\boldsymbol{\lambda}}_m , \tag{4.93}$$

so that (4.92) can be rewritten as

$$|\boldsymbol{x} - A_m \boldsymbol{\lambda}_m|^2 = |\widehat{\boldsymbol{v}}_m|^2 + (\widehat{\boldsymbol{\lambda}}_m - \boldsymbol{\lambda}_m)^{\top} D_m (\widehat{\boldsymbol{\lambda}}_m - \boldsymbol{\lambda}_m) . \tag{4.94}$$

Then consider that the likelihoods $L_m(\boldsymbol{x}, \boldsymbol{\lambda}_m)$ can be decomposed as follows

$$\begin{aligned} L_m(\boldsymbol{x}, \boldsymbol{\lambda}_m) &= \frac{1}{2\pi^{N}/2\sigma_{0m}^{N}} \exp\left\{-\frac{1}{2\sigma_{0m}^2}|\boldsymbol{x} - A_m \boldsymbol{\lambda}_m|^2\right\} = \\ &= \frac{\exp\left\{-\frac{1}{2\sigma_{0m}^2}|\widehat{\boldsymbol{v}}_m|^2\right\}}{2\pi^{\frac{N-N_m}{2}} \sigma_{0m}^{N-N_m} \sqrt{\det D_m}} \cdot \\ &\cdot \frac{\sqrt{\det D_m}}{(2\pi)^{N_m/2}\sigma_{0m}^{N_m}} \exp\left\{-\frac{1}{2\sigma_{0m}^2}(\widehat{\boldsymbol{\lambda}}_m - \boldsymbol{\lambda}_m)^{\top} D_m^{-1} (\widehat{\boldsymbol{\lambda}}_m - \boldsymbol{\lambda}_m)\right\} , \end{aligned} \tag{4.95}$$

where N is the dimension of $\boldsymbol{x}$ and N_m that of $\boldsymbol{\lambda}_m$.

As you can see, in (4.95) the first factor depends only on $\boldsymbol{x}$, while the second also depends on $\boldsymbol{\lambda}_m$. Moreover, the second factor is exactly a Gaussian, with mean $\widehat{\boldsymbol{\lambda}}_m$ and covariance $\sigma_{0m}^2 D_m^{-1}$; therefore its integral in $d\boldsymbol{\lambda}_m$ is then identically 1. Therefore we find

$$H_m(\boldsymbol{x}) = \int L(\boldsymbol{x}|\boldsymbol{\lambda}_m) d\boldsymbol{\lambda}_m = \frac{\exp\left\{-\frac{1}{2\sigma_{0m}^2}|\widehat{\boldsymbol{v}_m}|^2\right\}}{2\pi^{\frac{N-N_m}{2}} \sigma_{0m}^{N-N_m} \sqrt{\det D_m}} . \tag{4.96}$$

The maximum with respect to m of these expressions will indicate the most likely model. We note that the result coincides with that already known in the literature (see [4, 16, 44]).

For problem **(2)**: in this case it is sufficient to calculate

$$\begin{aligned} K_m(\boldsymbol{x}) - L_m(\boldsymbol{x}|\widehat{\boldsymbol{\lambda}}_m) &= \frac{1}{2\pi^{N/2}\sigma_{0m}^{N_m}} \exp\left\{-\frac{1}{2\sigma_{0m}^2}|\boldsymbol{x} - A_m \widehat{\boldsymbol{\lambda}}_m|^2\right\} = \\ &= \frac{1}{2\pi^{N/2}\sigma_{0m}^{N_m}} \exp\left\{-\frac{1}{2\sigma_{0m}^2}|\widehat{\boldsymbol{v}}_m|^2\right\} \end{aligned} \tag{4.97}$$

and then use (4.87) and (4.88) to find the optimal distribution of $\boldsymbol{x}$.

As you can see, the $K_m(\boldsymbol{x})$ are quite similar to the $H_m(\boldsymbol{x})$, but not equal precisely because the latter are obtained by integrating over $\boldsymbol{\lambda}_m$, while the former directly use the $\widehat{\boldsymbol{\lambda}}_m$ estimated model by model.

Chapter 5
Finite vs Infinite, Discrete vs Continuous

The subject of continuous vs. discrete modelling of physical phenomena is discussed and illustrated by an example taken from image analysis. The problem of estimating/predicting fields, i.e. infinite dimensional objects, from a finite set of observations is suitably defined. Deterministic solutions, with the two variants of the reduction of solution space or the Tikhonov regularization approach, are examined. The probabilistic, Bayesian, solution of the same problems is then undertaken, after presenting the basics of infinite dimensional probabilistic models.

5.1 Discrete and Finite Data, Continuous and Infinite Models; Observation Equations

The amount of data available to describe and interpret a certain phenomenon is always finite by nature. Even every real number that we use as data, in a calculation performed on a computer is actually represented by a finite sequence of zeros and ones.

It is good to remember that every real number, except for a negligible set, of measure zero, can only be expressed/represented with an infinite number of digits and operations. For example, if we use the standard decimal representation, almost all real numbers, in particular the irrationals, have a representation with infinite digits after the decimal point. Among other things, the representation of $\mathbb{R}$ given by J.W. R. Dedeking (last doctoral student of Gauss), as the set of contiguous classes of rational numbers, is essentially equivalent to that of completion of the field of rationals, with the ordinary Euclidean distance (*modulus of the difference*), therefore an operation based on the limit of an infinite sequence.

Given the topic we are discussing, it is worth remembering the definition of infinite set given by Dedekind: "infinite is a set for which it is possible to establish a

F. Sansò, A. Albertella, *The Probabilistic Vision of the Physical World*, Lecture Notes in Geosystems Mathematics and Computing,
https://doi.org/10.1007/978-3-031-88268-5_5

one-to-one correspondence with a proper subset of it". Therefore, no numerical data, storable and usable on a computer, can ever exactly contain an irrational number, but should we therefore give up the concept of real field? It would mean giving up the concept of limit, continuity, derivative, integral, and all the generalizations of functional spaces that follow from it; therefore we would not have the basic tools of mathematical analysis that allow us to identify the qualitative behavior of the solutions of many problems, which moreover could not even be expressed without those tools, for example in the form of differential equations.

If we think about it, there is a close, not occasional, analogy between data, as a number based on a finite number of digits, and real number, as a mathematical model, and empirical data versus theoretical model, derived by abstraction, which is the basis of knowledge as illustrated in the first chapter.

In a very similar way, the relationship between the data observable for a certain phenomenon and their mathematical representation, which typically includes the definition of some field, is posed.

A field $u(\boldsymbol{t})$ is essentially a function defined on a certain set that can be the time axis, or a portion of 3D space, or a variety in space, for example a sphere. $\boldsymbol{t}$ is a vector of coordinates that characterize the point at which u is defined. Furthermore, u in the simplest case is a function with real values, but it can also be a more complex object, for example a vector in $\mathbb{R}^N$, or a second order tensor or of a higher order.

For example, consider all observations based on the propagation of electromagnetic waves in the atmosphere: as is known, these are influenced by the refractive index, a field that will enter all observation equations.

Or, think of a magnetometer carried by an airplane: here the field, directly observed, is the magnetic induction vector. The same could be said of the gravity field, of which moreover with a gradiometer the tensor of second derivatives, the so-called *Marussi tensor*, can be observed.

Just as the "model" of a real number is that of belonging to the field $\mathbb{R}$, satisfying its definitions, so the "model" of a field is that of being a member of a functional space H. For simplicity, we assume that $u(\boldsymbol{t})$ is just a function and that the space H to which it belongs is a Hilbert space, although in many cases it would be more appropriate to think of more general spaces. We recall that a Hilbert space is by definition equipped with a scalar product, $< u, v >$, between two elements u and v, and that these, when their product is zero, are said to be orthogonal to each other.

Another simplification that we assume as a hypothesis is that the space H is separable, which is a reasonable assumption in almost all concrete cases. This is equivalent to stating that there exists in the space a sequence of functions $\{h_n(\boldsymbol{t})\}$ that is orthogonal and complete. This means that the relation holds

$$< h_n(\boldsymbol{t}), h_m(\boldsymbol{t}) >= \delta_{nm} \tag{5.1}$$

and also that every element $u \in H$ can be developed in the generalized Fourier series,

$$u(t) = \sum_{n=1}^{+\infty} \xi_n h_n(t) \; . \tag{5.2}$$

The convergence of the series (5.2) must be understood in the sense of the norm of H, induced by the scalar product $<,>$ according to the rule

$$\| u - v \|^2 = < u - v, u - v > \; . \tag{5.3}$$

We still remember that in this hypothesis the remarkable Parseval theorem holds, which states that

$$\| u \|^2 = \sum_{n=1}^{+\infty} \xi_n^2 \tag{5.4}$$

as is easy to deduce from (5.3).

The (5.4) directly demonstrates that, once the orthonormal basis $\{h_n(t)\}$ is chosen, there is an isometric isomorphism between H and the Landau space ℓ^2, that is the space of real sequences $\{a_n\}$ equipped with scalar product

$$(\{a_n\} \, , \, \{b_n\})_{\ell^2} = \sum_{n=1}^{+\infty} a_n b_n \tag{5.5}$$

and norm

$$\| \{a_n\} \|_{\ell^2} = \sqrt{\sum_{n=1}^{+\infty} a_n^2} \; . \tag{5.6}$$

In practice we can say that every separable space H is directly representable as ℓ^2, which in turn is a natural infinite-dimensional generalization of the Euclidean space $\mathbb{R}^N$.

At the level of notation it is useful for us to generalize the algebraic one, used for Euclidean spaces, for example by setting

$$\boldsymbol{\xi} = \{\xi_n \, , \, n = 1, 2, \ldots\} \, , \; \boldsymbol{h}(t) = \{h_n(t) \, , \, n = 1, 2 \ldots\} \, , \tag{5.7}$$

$$u(t) = \boldsymbol{\xi}^\top \boldsymbol{h}(t) \equiv \boldsymbol{h}(t)^\top \boldsymbol{\xi} \, , \tag{5.8}$$

$$\| u \|^2 = |\boldsymbol{\xi}|^2 \equiv \boldsymbol{\xi}^\top \boldsymbol{\xi} \, , \tag{5.9}$$

$$u = \boldsymbol{\xi}^\top \boldsymbol{h}(t) \, , \; v = \boldsymbol{\eta}^\top \boldsymbol{h}(t) \Rightarrow < u, v > = \boldsymbol{\xi}^\top \boldsymbol{\eta} \; . \tag{5.10}$$

Now, we need to establish the relationship between the empirical data of a certain experiment, for example of measurement, and the "model" of field $u(t)$. This relationship is provided by a physical law that represents the operation of the measurement to the best of our knowledge. If the output of the observation is a real number, X, we will write that

$$X = F[u(\boldsymbol{t}), \boldsymbol{\lambda}] + \nu\,; \tag{5.11}$$

F is generally a non-linear functional of $u(\boldsymbol{t})$, which often contains a certain number of auxiliary parameters, collected in the vector $\boldsymbol{\lambda}$, ν is a discrepancy between the law represented by F and the numerical value of X, which we call measurement error.

It is interesting to note that the nature of ν can be or not be in principle stochastic, since in ν various types of error converge, including the errors of the model represented by F, [11, 52]. However, here we limit ourselves to the case that the various uncontrolled disturbance factors from the analytical expression $F(u, \boldsymbol{\lambda})$ combine to give rise to a random variable ν which we will assume to be zero mean. Finally, in order to avoid to face greater mathematical complexity, we will assume that in (5.11) F is linearizable, or that $u(\boldsymbol{t})$ can be considered small compared to a fixed and already known model. This happens for example for the potential of the gravity field, in which the part of anomalous potential, outside the masses, is $\sim 10^{-5}$ times that of the normal potential, which is a well-known function.

Lastly, wanting to focus on the relationship between the vector of available observations $\boldsymbol{X}$ and the unknown field u, we assume that this does not contain ancillary parameters $\boldsymbol{\lambda}$; essentially we come to study the simplified model

$$\boldsymbol{X} = \boldsymbol{L}u(\boldsymbol{t}) + \boldsymbol{\nu} \tag{5.12}$$

in which $\boldsymbol{X}$ is the vector of observables, $\boldsymbol{\nu}$ the vector of measurement errors for which it is assumed that

$$E\{\boldsymbol{\nu}\} = 0\,,\ \ E\{\boldsymbol{\nu}\boldsymbol{\nu}^{\top}\} = C_{\nu}\,, \tag{5.13}$$

with C_ν known covariance matrix, and finally $\boldsymbol{L}$ is a vector of functionals

$$\boldsymbol{L} = \begin{vmatrix} L_1 \\ \dots \\ L_m \end{vmatrix} \tag{5.14}$$

so that

$$\boldsymbol{L}u(\boldsymbol{t}) \equiv \begin{vmatrix} L_1[u(\boldsymbol{t})] \\ \dots \\ L_m[u(\boldsymbol{t})] \end{vmatrix}\,. \tag{5.15}$$

It is important to observe already here that for a certain functional L to represent an observable quantity, L must be a continuous functional on H.

Therefore, in our scheme (5.12) the data are a finite (and obviously discrete) set, while the unknown $u(\boldsymbol{t})$ is a variable that has the power of the continuum and infinite degrees of freedom. The vector $\boldsymbol{\nu}$ will always be a r.v. in $\mathbb{R}^m$ for us, and so $\boldsymbol{X}$; $u(\boldsymbol{t})$ on the contrary may or may not have a stochastic modeling depending on whether the estimation problem—for a field however the term prediction is used—is approached from a frequentist point of view, for which $u(\boldsymbol{t})$ is a deterministic variable, or Bayesian, for which $u(\boldsymbol{t})$ becomes a random field (r.f.).

On the other hand, it is important to note that the qualification discrete-finite vs continuous-infinite of a model should never be exclusive. For reasons of computability, we must be able to represent the field $u(\boldsymbol{t})$ by a finite number of degrees of freedom, naturally with the guarantee that when these are made to grow to infinity, the finite scheme approximates $u(\boldsymbol{t})$ in H. On the contrary, when the observations $\boldsymbol{X}$ concern point values of the field, or some of its functional, then it can be useful to switch to a continuous description of the observations because this often allows to capture qualitative characteristics of the solution, which purely numerical estimates fail to highlight.

It should be noted that in this case also ν must become a stochastic field of noise; at this level there are more choices but perhaps the simplest is to model ν as a Wiener measure, as well illustrated in mathematical, geodetic and geophysical literature, [23, 43].

It is useful to give a criterion, at least in principle, to decide when the transition from a discrete model to a continuous one or vice versa can give compatible results. A "golden rule" can thus be stated [43]: suppose that $\boldsymbol{X}$ represents point-type measurements and let d_0 be a characteristic distance between the measurement points; let $\gamma(d)$ be the variogram of the field $u(\boldsymbol{t})$, or the empirical average of the variation $|u(\boldsymbol{t}) - u(\boldsymbol{t} + \boldsymbol{\tau})|^2$ when $|\boldsymbol{\tau}| = d$; then the representation of $\boldsymbol{X}$ as a continuous field is reasonable if

$$\gamma(d_0) \ll \sigma_\nu^2 \,. \tag{5.16}$$

It is clear that, if (5.16) holds, the discrete values can be interpolated with any smooth interpolator, without introducing significant errors with respect to ν. The same criterion can be used in reverse, if we want to represent a continuous field, for example through its values on a grid of points.

An example from image analysis helps to clarify the concepts presented.

Example 5.1 A grayscale image is a vector $\boldsymbol{X}$ of a given function on a regular grid that covers a rectangle in the plane. The rectangles of the grid that have centers with integer coordinates (i, k), are called pixels, π_{ik}, (see [13]). The single observation X_{ik} represents the flow of electromagnetic energy, through π_{ik}, possibly related to a certain frequency channel, in remote sensing problems; this energy, accumulated over the measurement time, has a surface density that represents our unknown field.

$$u(t_1, t_2) = u(\boldsymbol{t}) \,, \tag{5.17}$$

with (t_1, t_2) Cartesian coordinates on the image plane. The observation equation is therefore

$$X_{ik} = \int_{\pi_{ik}} u(t)dt + \nu_{ik} \,. \tag{5.18}$$

The noise ν_{ik} is generated by the electronic measurement apparatus. Therefore, typically (5.18) is a system of discrete and finite observation equations, while the unknown, once the space H to which it must belong has been fixed, is a function with infinite degrees of freedom. A standard problem in image analysis is to estimate a sufficiently smooth $u(\boldsymbol{t})$, lowering the effect of the noise, the so-called *denoising*, [13]. It is not the case here to delve into the problem of identifying the discontinuities that typically appear in an image. Here we just want to underline that, in the event that the original image is substantially smooth, the original unknown $u(\boldsymbol{t})$ could be replaced by a grid of values interpolated by bicubic splines, so that the unknown would become finite-dimensional. Indeed, if the (5.16) were verified on the distance of some pixels, the grid of the unknowns could be a sub-grid of the observations and the problem would thus fall back into the scope of a least squares estimate. Always under the same condition, one could instead transform the vector of $\boldsymbol{X}$ into a continuous field $X(\boldsymbol{t})$, and in this case one could analytically express the solution $\widehat{u}(\boldsymbol{t})$, given by a Wiener filter. The conceptual difficulty in this case is rather that of handling a continuous noise $\nu(\boldsymbol{t})$ that must be represented by a random Wiener measure $\nu(\boldsymbol{t})$, characterized by the relations

$$E\{\nu(dt)\} = 0 \,, \; E\{\nu(dt)\nu(dt')\} = \sigma_0^2 \mu(dt \cap dt') \,, \tag{5.19}$$

with μ the ordinary measure in $\mathbb{R}^2$ (see [42]).

The original discrete noise can then be seen as a discretization of the $\nu(dt)$, characterized by the relation

$$\nu_{ik} = \int_{\pi_{ik}} \nu(dt) \,, \tag{5.20}$$

or, in terms of variance,

$$\sigma^2(\nu_{ik}) = \sigma_0^2 \mu(\pi_{ik}) \,. \tag{5.21}$$

5.2 Deterministic Solutions

A bibliographic reference for what is discussed in this section can be found in [47]. We now resume the original scheme

$$\boldsymbol{X} = \boldsymbol{L}(u(\boldsymbol{t})) + \boldsymbol{\nu} \,; \tag{5.22}$$

with the condition that $u \in H$ we can use the (5.8) to rewrite the (5.22) in the form

$$X = L(h(t)^\top)\xi + \nu \equiv M\xi + \nu \,, \tag{5.23}$$

where ξ is an infinite-dimensional vector, $\xi \in \ell^2$, and M a matrix of m rows and infinite columns,

$$M_{ik} = L_i(h_k(t)) \,;$$

note that for the product between a row of M and any $\xi \in \ell^2$ to be finite, it is also necessary that each row $m_i^\top \equiv \{M_{ik}\ k = 1 \ldots\}$ is in ℓ^2.

Written in the form (5.22), without the vector of residuals ν, it is clear that the problem of determining ξ from X is underdetermined, i.e. the unknowns are an infinite vector while the data are finite. There are two main solutions used for this purpose: the reduction of the solution space or the regularization approach.

In the case of reducing the solution space, the idea is that, especially if the base functions $h_i(t)$ for high values of i are oscillating with increasing frequency, mindful of the Nyquist theorem [37], one can think that the components of ξ from a certain point onwards are irrelevant.

In other words, given

$$\xi_n^\top = [\xi_1 \ldots \xi_n \,, 0, 0 \ldots] \tag{5.24}$$

and

$$(\xi^n)^\top = [0 \ldots 0, \xi_{n+1}, \ldots] \tag{5.25}$$

one can suppose that $(\xi^n)^\top h(t)$ has a small impact on the observables X.

One can then replace the condition $u \in H$ with

$$u \in H_n \equiv \{u = h^\top(t)\xi_n\} \,; \tag{5.26}$$

that is, u is constrained to belong to the reduced space H_n instead of H.

If we then set

$$\begin{cases} M \equiv [M_n \,,\ M^n] \\ M_n = [M_{ik} \,;\ i = 1 \ldots m, k = 1 \ldots n] \\ M^n = [M_{ik} \,;\ i = 1 \ldots m, k = n+1 \ldots] \end{cases} \tag{5.27}$$

Eq. (5.23) can be rewritten in an approximate form, with an obvious abuse of notation, as

$$X = M_n \xi_n + \nu \,, \tag{5.28}$$

and when $n < m$, the problem becomes over-determined again and its solution becomes classical, for example that of least squares. Of course, it is interesting to understand what is the estimation error of ξ, and therefore of u, when we set $\widehat{\xi} = \widehat{\xi}_n$, also to clarify under what conditions the approximation made is acceptable.

Using the isometry (5.9) and the fact that ξ_n and ξ^n are orthogonal to each other, the total estimation error can be expressed as

$$\begin{aligned} \mathcal{E}^2 &= E\{\| u - \widehat{u} \|^2\} = E\{|\xi - \widehat{\xi}|^2\} = \\ &= E\{|\xi_n - \widehat{\xi}_n|^2 + |\xi^n|^2\} = E\{|\xi_n - \widehat{\xi}_n|^2\} + |\xi^n|^2 \end{aligned} \tag{5.29}$$

The term $|\xi^n|^2$ is called *omission error*, (see [47], Chap. 3)

$$\mathcal{O}^2 = |\xi^n|^2 = \sum_{k=n+1}^{+\infty} \xi_k^2 , \tag{5.30}$$

and it is clear that it is a decreasing function of n. The term $E\{|\xi_n - \widehat{\xi}_n|^2\}$ is called *commission error* $\mathcal{C}^2$ and requires a more refined analysis. Let's set

$$\widehat{\xi}_n = S_n X \ , \ S_n = (M_n^\top C_\nu^{-1} M_n)^{-1} M_n^\top C_\nu^{-1} \ , \tag{5.31}$$

and note that $S_n M_n = I$, and also

$$X = M_n \xi_n + M^n \xi^n + \nu \ , \tag{5.32}$$

so that

$$\widehat{\xi}_n = \xi_n + S_n M^n \xi^n + S_n \nu \ . \tag{5.33}$$

Therefore it results

$$\xi_n - \widehat{\xi}_n = -S_n M^n \xi^n - S_n \nu \ ; \tag{5.34}$$

Since ν is zero mean by hypothesis, it is clear that $\xi_n - \widehat{\xi}_n$ has a bias b_n,

$$b_n = -S_n M^n \xi^n \ . \tag{5.35}$$

We can therefore conclude that the commission error results

$$\mathcal{C}^2 = E\{|\xi_n - \widehat{\xi}_n|^2\} = |b_n|^2 + Tr(M_n^\top C_\nu^{-1} M_n)^{-1} \ . \tag{5.36}$$

The bias term

$$|b_n|^2 = (\xi^n)^\top (M^n)^\top S_n^\top S_n M^n \xi^n \tag{5.37}$$

has a behavior as a function of n that must be examined case by case, even if it is plausible that $|b_n|^2$ decreases at least from a certain n onwards, since $\xi^n \to 0$.

On the contrary, it is possible to see, using the formula for the inversion of a block matrix, that $Tr(M_n^\top C_\nu^{-1} M_n)^{-1}$ is always increasing with n. This shows that in any case the total error (5.29) will have a minimum; if this happens with a value less than or equal to m, we will have determined an optimal choice of n.

In any case, the approach of reducing the solution space combined with the least squares estimate becomes acceptable when it results

$$|\boldsymbol{b}_n|^2 + |\boldsymbol{\xi}^n|^2 < Tr(M_n^\top C^{-1} M_n)^{-1} . \tag{5.38}$$

It is clear that a verification of (5.38) necessarily requires that we have some prior knowledge of the behavior of $\boldsymbol{\xi}^n$ as a function of n. It is exactly this kind of need that pushes us to consider as useful the Bayesian approach, as we will see later.

Example 5.2 A notable example of the use of solution space reduction is that of determining a global model of the anomalous potential of the Earth's gravity from given gravity anomaly data on the surface. We present here an extremely simplified version. For a more in-depth discussion, see [47] chapters 1, 7.

Suppose that the "observed" data are values of gravity anomalies Δg on an equi-angular grid of side δ on the Earth's sphere, with radius $R = 6371$ km. Taken spherical coordinates $\sigma \equiv (\vartheta, \lambda), 0 \leq \vartheta \leq \pi, 0 \leq \lambda \leq 2\pi$, the points of the grid, with $\delta = \frac{\pi}{N}$, when we skip the poles, will be

$$\begin{cases} \sigma_{ik} = (\vartheta_i, \lambda_k) & \\ \vartheta_i = i\delta & i = 1 \ldots N-1 \\ \lambda_k = k\delta & k = 1 \ldots 2N \end{cases} \tag{5.39}$$

that is, the number of data is

$$m = 2N(N-1) . \tag{5.40}$$

The unknown here is the anomalous field $T(r, \sigma)$ which for a spherical domain, $\{r \geq R\}$, can be represented by the series

$$T(r, \sigma) = \frac{\mu}{R} \sum_{j=2}^{+\infty} \sum_{\ell=-n}^{n} T_{j\ell} \left(\frac{R}{r}\right)^{j+1} Y_{j\ell}(\sigma) , \tag{5.41}$$

($\mu = GM_t$, G is Newton's constant, M_t is the mass of the Earth)

where $Y_{nm}(\sigma)$ are the spherical harmonics (see [47], Chap. 3), which represent an orthonormal and complete system in $L^2(\sigma)$, satisfying the condition

$$\frac{1}{4\pi} \int_\sigma Y_{rs}(\sigma) Y_{j\ell}(\sigma) d\sigma = \delta_{rj} \delta_{s\ell} . \tag{5.42}$$

$$(d\sigma = \sin\vartheta d\vartheta d\lambda)$$

The series (5.41) starts from $j = 2$ because the coefficient T_{00} is known and the $T_{1\ell}$ are null by convention (see [47], Chap. 3). The series is convergent for $r \geq R$, and in particular on the sphere $r = R$ we have

$$T(R, \sigma) = \frac{\mu}{R} \sum_{j=2}^{+\infty} \sum_{\ell=-j}^{j} T_{j\ell} Y_{j\ell}(\sigma) \tag{5.43}$$

The gravity anomalies Δg are related to T by the relation

$$\Delta g(r, \sigma) = \frac{\mu}{R^2} \sum_{j=2}^{+\infty} \sum_{\ell=-j}^{j} (j-1) T_{j\ell} \left(\frac{R}{r}\right)^{j+2} Y_{j\ell}(\sigma) \tag{5.44}$$

which on the sphere becomes

$$\Delta g(R, \sigma) = \frac{\mu}{R^2} \sum_{j=2}^{+\infty} \sum_{\ell=-j}^{j} (j-1) T_{j\ell} Y_{j\ell}(\sigma) \,. \tag{5.45}$$

To ensure that the (5.45) is uniformly convergent it is convenient to assume that the space H to which T must belong is such that

$$\sum_{j=2}^{+\infty} \sum_{\ell=-j}^{j} (j+1)^3 T_{j\ell}^2 < +\infty \,. \tag{5.46}$$

Therefore, the observation equations for the grid (5.39) are

$$\Delta g_0(\sigma_{ik}) = \frac{\mu}{R^2} \sum_{j=2}^{+\infty} \sum_{\ell=-j}^{j} (j-1) T_{j\ell} Y_{j\ell}(\sigma_{ik}) + \nu_{ik} \,, \tag{5.47}$$

where the index 0 in $\Delta g_0(\sigma_{ik})$ underlines that these are observed quantities.

If the values $\Delta g_0(\sigma_{ik})$ are derived from averages of point values on an area surrounding the node σ_{ik}, if we suppose that the observations are independent and that the individual points are uniformly distributed on the sphere, it is not unreasonable to assume that

$$\sigma^2(\nu_{ik}) = \frac{\sigma_0^2}{\sin \vartheta_i} \,; \tag{5.48}$$

in fact if we stipulate the ν_{ik} to be uncorrelated with each other, the variance of the average will be proportional to the inverse of the number of points over which it was taken, and this in turn is proportional to the area, which is approximately $(\sin \vartheta_i \delta^2)$.

In the observation equations $\Delta g_0(\sigma_{ik})$ are the components of the vector $\boldsymbol{X}$, which as we have seen are $2N(N-1)$. The unknown parameters T_{ij} are the components of the vector $\boldsymbol{\xi}$. The reduction of the solution space consists in limiting the degree j according to the relation

$$j \leq L \,. \tag{5.49}$$

In this case the number of parameters is $n = \sum_{j=2}^{L}(2j+1) = (L+1)^2 - 4$, so for each given N, if it is

$$L < \sqrt{2N(N-1)+4} - 1 \tag{5.50}$$

the problem is over-determined. Therefore, the design matrix of the measurements in this case is

$$M_{(ik),(j\ell)} = \left(\frac{\mu}{R^2}\right)(j-1)Y_{j\ell}(\sigma_{ik}) \tag{5.51}$$

and the normal matrix is

$$(M^\top C_\nu^{-1} M)_{(rs),(j\ell)} = \left(\frac{\mu}{R^2}\right)^2 (r-1)(j-1)\sum_{i,k} Y_{rs}(\sigma_{ik})Y_{j\ell}(\sigma_{ik})\frac{\sin\vartheta_i}{\sigma_0^2} \,. \tag{5.52}$$

Using the following approximation of the (5.42), obtained by discretization,

$$\delta_{jr}\delta_{\ell s} = \frac{1}{4\pi}\int Y_{j\ell}(\sigma)Y_{rs}(\sigma)d\sigma \cong \frac{1}{4\pi}\sum_{i,k} Y_{j\ell}(\sigma_{ik})Y_{rs}(\sigma_{ik})\sin\vartheta_i\delta^2 \,, \tag{5.53}$$

we see that the (5.52) can be written as

$$(M^\top C_\nu^{-1} M)_{(rs),(j\ell)} = \frac{4\pi}{\sigma_0^2\delta^2}(j-1)^2\left(\frac{\mu}{R^2}\right)^2 \delta_{rj}\delta_{\ell s} \,, \tag{5.54}$$

that is, the matrix, in the given approximation, is essentially diagonal.

If we then consider the normal known term

$$(M^\top C_\nu^{-1} \boldsymbol{X})_{(j\ell)} = \left(\frac{\mu}{R^2}\right)(j-1)\sum_{i,k}\Delta g_0(\sigma_{ik})Y_{j\ell}(\sigma_{ik})\frac{\sin\vartheta_i}{\sigma_0^2} \tag{5.55}$$

we see that it is possible to explicitly write the solution as

$$\widehat{T}_{j\ell} = \left(\frac{\mu}{R^2}\right)^{-1}\frac{1}{(j-1)}\frac{\delta^2}{4\pi}\sum_{i,k}\Delta g_0(\sigma_{ik})Y_{j\ell}(\sigma_{ik})\sin\vartheta_i \,. \tag{5.56}$$

The elaboration can be pushed further, with some approximation, to analytically calculate the commission error $\mathcal{C}^2$. Therefore, assuming a certain trend for $\mathcal{O}^2$, it is possible to find the optimal value of L, even though this more technical topic is beyond the scope of this discussion.

The regularization approach is actually a discrete form for the measurement part, of the principle already studied in functional terms by Arsenin and Tikhonov for the solution of ill-posed problems [25, 54] . Tikhonov's approach to solving an ill-posed linear problem (or better a simplified version of it),

$$X = Au \ , \ \ X \in \mathcal{K}, u \in H \ , \tag{5.57}$$

with $\mathcal{K}$ and H suitable spaces, for example Hilbert spaces, is to seek the minimum of the functional

$$F(u, \lambda) \equiv \| X - Au \|_{\mathcal{K}} + \lambda < u, Qu >_H \ , \tag{5.58}$$

where Q is a strictly positive operator in H, and in which the aim is to balance the approximation of X with Au while at the same time keeping u smooth, i.e., with a not too large norm. In the simplest case, the choice of Q is $Q = I$, and therefore the regularization term is precisely $\| u \|_H^2$. The parameter λ, which weighs the relative importance of the two terms in (5.58), will be discussed later. Here we are interested in seeing how (5.58) should be modified when X, instead of being a function, is an observation vector like that of our model

$$\boldsymbol{X} = M\boldsymbol{\xi} + \boldsymbol{\nu} \ , \tag{5.59}$$

in which we recall that $\boldsymbol{\xi}$ must be in ℓ^2 if we want $u \in H$.

If we tried to apply the least squares principle to (5.23) we would obtain from the minimum condition the usual normal system

$$(M^\top C_\nu^{-1} M)\boldsymbol{\xi} = M^\top C_\nu^{-1} \boldsymbol{X} \ . \tag{5.60}$$

If we suppose that the rows $\boldsymbol{m}_i^\top$ of M, or the columns $\boldsymbol{m}_i \in \ell^2$ of $M^\top$, are linearly independent, it is easy to see that (5.60) always admits a unique solution in the subspace

$$H_m = span M^\top \equiv \{M^\top \boldsymbol{\tau} \ , \ \boldsymbol{\tau} \in \mathbb{R}^m\} \equiv span\{\boldsymbol{m}_i \ , \ i = 1 \dots m\} \ . \tag{5.61}$$

In fact, setting $\boldsymbol{\xi} = M^\top \boldsymbol{\tau}$ in (5.60) it will be enough to solve the isodetermined system

$$MM^\top \boldsymbol{\tau} = \boldsymbol{X} \ ; \tag{5.62}$$

note that $MM^\top$ is a square matrix of dimension m, symmetric and positive definite. Naturally, (5.62) does not have a unique solution; any ξ_0 for which

$$M\xi_0 = 0 \Rightarrow \xi_0 \perp span\{\boldsymbol{m}_i\} \tag{5.63}$$

can be added to $M^\top \boldsymbol{\tau}$, found above, so that $\xi_0 + M\tau$ is a general solution of the normal system. Naturally, there are infinitely many $\boldsymbol{\xi}_0$ as in (5.63), all vectors in ℓ^2 in the space complementary to $spanM^\top$. Here lies the improper posedness of our problem and therefore a possible recourse to the principle of Arsenin-Tikhonov [54].

So we look for the solution $\boldsymbol{\xi}$ that minimizes

$$F(\boldsymbol{\xi}, \lambda) = (\boldsymbol{X} - M\boldsymbol{\xi})^\top C_\nu^{-1}(\boldsymbol{X} - M\boldsymbol{\xi}) + \lambda\boldsymbol{\xi}^\top Q\boldsymbol{\xi} \tag{5.64}$$

with Q strictly positive definite in ℓ^2. The minimum condition of $F(\boldsymbol{\xi}, \lambda)$ leads to the system

$$(M^\top C_\nu^{-1} M + \lambda Q)\boldsymbol{\xi} = M^\top C_\nu^{-1}\boldsymbol{X}\,. \tag{5.65}$$

It is clear that the operator $M^\top C_\nu M$ is not invertible in ℓ^2, but the addition of the positive operator Q makes the system (5.65) solvable and we can safely write

$$\widehat{\boldsymbol{\xi}} = (M^\top C_\nu^{-1} M + \lambda Q)^{-1} M^\top C_\nu^{-1}\boldsymbol{X}\,. \tag{5.66}$$

When the choice $Q = I$ is made, the estimator (5.66) is called in statistical literature *ridge regression estimator* [7].

The estimator thus found is generally biased, in fact, using (5.59) in (5.66),

$$\begin{aligned} \boldsymbol{b}(\lambda) &= E\{\widehat{\boldsymbol{\xi}}\} - \boldsymbol{\xi} = (M^\top C_\nu^{-1} M + \lambda Q)^{-1} M^\top C_\nu^{-1} M\boldsymbol{\xi} - \boldsymbol{\xi} \\ &= -\lambda(M^\top C_\nu^{-1} M + \lambda Q)^{-1} Q\boldsymbol{\xi}\,. \end{aligned} \tag{5.67}$$

Furthermore, the random part of the error in $\widehat{\boldsymbol{\xi}}$ is

$$\boldsymbol{\varepsilon} = (M^\top C_\nu^{-1} M + \lambda Q)^{-1} M^\top C_\nu^{-1}\boldsymbol{\nu}\,. \tag{5.68}$$

Therefore, the total error of $\widehat{\boldsymbol{\xi}}$ is

$$\begin{aligned} \mathcal{E}_t^2(\lambda) &= |\boldsymbol{b}(\lambda)|^2 + E\{|\boldsymbol{\varepsilon}(\lambda)|^2\} \equiv |\boldsymbol{b}(\lambda|^2 + \mathcal{C}^2(\lambda) = \\ &= \lambda^2\boldsymbol{\xi}^\top Q(M^\top C_\nu^{-1} M + \lambda Q)^{-2} Q\boldsymbol{\xi} + \\ &+ Tr(M^\top C_\nu^{-1} M)(M^\top C_\nu^{-1} M + \lambda Q)^{-2}\,. \end{aligned} \tag{5.69}$$

Naturally, expression (5.69) encourages the search for a value of λ that minimizes the total error. The method was developed by Morozov, see [46] App. C for a detailed analysis.

Here we will limit ourselves to showing that one can expect a minimum of $\mathcal{E}_t^2(\lambda)$ as $|\boldsymbol{b}(\lambda)|^2$ is an increasing function of λ, while $\mathcal{C}^2(\lambda) = E\{|\boldsymbol{\varepsilon}(\lambda)|^2\}$ is a decreasing function. We see this under the simplest assumption that $Q = I$.

As already seen in (5.61), the range of $M^\top$ is the m-dimensional space H_m; this is then also the range of $M^\top C_\nu^{-1} M$. Therefore, the symmetric normal operator $M^\top C_\nu M$ will only have m positive eigenvalues, while the entire subspace $H_m^\perp$ will correspond to the null eigenvalue. Then, there exists an orthonormal basis $\{\boldsymbol{e}_i\}$ of H_m, corresponding to m eigenvalues of $\lambda_i > 0$, which allows the normal operator to be represented as

$$M^\top C_\nu^{-1} M = \sum_{i=1}^{m} \lambda_i \boldsymbol{e}_i \boldsymbol{e}_i^\top . \tag{5.70}$$

We complete the basis $\{\boldsymbol{e}_i\}$ with orthonormal vectors that span the space $H_m^\perp$, so returning to (5.67), setting $Q = I$, we find

$$|\boldsymbol{b}(\lambda)|^2 = \lambda^2 \sum_{i=1}^{+\infty} \frac{(\boldsymbol{e}_i^\top \boldsymbol{\xi})^2}{(\lambda_i + \lambda)^2} = \lambda^2 \sum_{i=1}^{m} \frac{(\boldsymbol{e}_i^\top \boldsymbol{\xi})^2}{(\lambda_i + \lambda)^2} + \sum_{i=m+1}^{+\infty} (\boldsymbol{e}_i^\top \boldsymbol{\xi})^2 \tag{5.71}$$

where it is taken into account that $\lambda_i = 0$ for $i > m$. Since the function $\lambda^2/(\lambda_i + \lambda)^2$ is increasing in λ, so is $|b(\lambda)|^2$.

More precisely, setting $P_m^\perp$ the orthogonal projector on $H_m^\perp$, we see $|\boldsymbol{b}(\lambda)|^2$ to go from $|P_m^\perp \boldsymbol{\xi}|^2$, for $\lambda \to 0$, to $|\boldsymbol{\xi}|^2$, for $\lambda \to +\infty$. As for the second term of (5.69), we have, using the spectral representation (5.70),

$$\mathcal{C}^2(\lambda) = E\{|\boldsymbol{\varepsilon}(\lambda)|^2\} = \sum_{i=1}^{m} \frac{\lambda_i}{(\lambda_i + \lambda)^2} . \tag{5.72}$$

Naturally, the sum (5.72) is limited to m because $\lambda_i = 0$ for $i > m$. As you can see, this function is decreasing in λ, and in particular it will be

$$\mathcal{C}^2(0) = \sum_{i=1}^{m} \frac{\lambda_i}{\lambda_i^2}, \text{ while } \mathcal{C}^2(\lambda) \to 0 \text{ for } \lambda \to +\infty . \tag{5.73}$$

Example 5.3 We revisit here the case of Example 5.2, with the same observation equations of a grid of gravity anomalies $\Delta g_0(\sigma_{ik})$, this time seeking a regularized solution. The normal matrix is the same (5.54), that is diagonal, to which however λI will have to be added if a ridge regression solution has been chosen. Just to make the effect of the constant λ more evident, we can think of replacing this with a parameter γ defined by the relation

$$\lambda = \frac{4\pi}{\sigma_0^2 \delta^2} \left(\frac{\mu}{R^2} \right)^2 \gamma \, , \tag{5.74}$$

and arrive at writing the solution

$$\widehat{T}_{j\ell} = \left(\frac{\mu}{R^2} \right)^{-1} \frac{\delta^2}{4\pi} \frac{j-1}{(j-1)^2 + \gamma} \cdot \sum_{ik} \Delta g_0(\sigma_{ik}) Y_{j\ell}(\sigma_{ik}) \sin \vartheta_i \, . \tag{5.75}$$

For future comparison reasons, it is convenient to write the most general solution when choosing for Q, instead of I, a diagonal matrix with increasing elements q_j. If we put, with obvious symbolism,

$$\Delta g_{0j\ell} = \frac{1}{4\pi} \sum_{ik} \Delta g_0(\sigma_{ik}) Y_{j\ell}(\sigma_{ik}) \sin \vartheta_i \delta^2 \, , \tag{5.76}$$

the three solutions so far elaborated are written

$$\widehat{T}_{j\ell} = \left(\frac{\mu}{R^2} \right)^{-1} \frac{1}{j-1} \Delta g_{0j\ell} \tag{5.77}$$

$$\widehat{T}_{j\ell} = \left(\frac{\mu}{R^2} \right)^{-1} \frac{j-1}{(j-1)^2 + \gamma} \Delta g_{0j\ell} \tag{5.78}$$

$$\widehat{T}_{j\ell} = \left(\frac{\mu}{R^2} \right)^{-1} \frac{j-1}{(j-1)^2 + \gamma q_j^2} \Delta g_{0j\ell} \, . \tag{5.79}$$

5.3 The Bayesian Approach

In order to generally address the problem of using observation equations (5.12) to obtain an estimate of the unobservable variable $u(t) \in H$, it is first necessary to understand how to make $u(t)$ a random variable that has its own distribution in H, that is, in an infinite-dimensional space.

It is important to understand that here $\{u(t)\}$ should be thought of as the set of all values that define the function $u(t)$, therefore as a single point in H.

A further complication in performing probabilistic calculations in an infinite-dimensional space is that we do not have the concept of probability density in the usual sense, since in such spaces a Lebesgue measure, that is, a translation invariant measure, with respect to which the density is defined, cannot be defined. In most texts on random fields, these difficulties are solved by the Kolmogorov approach, which defines the random variables $u(t, \omega)$ with t fixed and ω belonging to a fixed probability space $(\Omega, \mathcal{A}, P)$ and starts from the distribution of the N-tuples of random variables $\{u(t_1) \dots u(t_N)\}$, considered as marginal distributions of a distribution in the space of all functions on T, and then studies under what conditions it happens that $u(t, \omega) \in H$ with $P = 1$.

However, this approach has two logical difficulties: the first is that the probability space Ω is not explicitly defined, the second is that for many Hilbert spaces the point value of the function $u(t, \omega)$ is not defined; for example, if H is $L^2(T)$ the evaluation functional in t of u, $u(t, \omega)$, is notoriously not continuous.

However, the Kolmogorov approach is not the only one (see [16], Chap. I) and here we will follow a path that is simpler, albeit less general. Taking advantage of the isometry (5.9), $(u \in H)) \Rightarrow (\boldsymbol{\xi} \in \ell^2)$, we start by defining a distribution for any sequence $\{\boldsymbol{\xi}\} \in R^\infty$.

In this case, the Kolmogorov theorem (see [16], Chap. I, [44], Chap. 8) takes a much simpler form.

Let $\mathbb{R}^n \subset \mathbb{R}^{n+1} \dots \subset \mathbb{R}^\infty$ be the sequence of subspaces of $\mathbb{R}^\infty$, each of finite dimension and contained in the next, in the sense that

$$\boldsymbol{\xi}^{n+1} \in \mathbb{R}^{n+1} \equiv \{\boldsymbol{\xi}^n \in \mathbb{R}^n \ , \ \forall \xi_{n+1}\} \, . \tag{5.80}$$

Let P^n, $P^{n+1} \dots$ also be the orthogonal projectors of $\mathbb{R}^\infty$ onto $\mathbb{R}^n$, $\mathbb{R}^{n+1}$ in the sense that

$$\boldsymbol{\xi} \in \mathbb{R}^\infty \ , \ P^n \xi = [\xi_1, \dots \xi_n] \, . \tag{5.81}$$

We define a cylinder C_n of $\mathbb{R}^\infty$, with base the Borel set $B_n \subset \mathbb{R}^n$, the set

$$C_n = \{\boldsymbol{\xi} \ ; \ P^n \xi \in B_n\} \, ; \tag{5.82}$$

it is easy to see that the family of cylinders,

$$\mathcal{C} \equiv \{C_n \ ; \ \forall B_n \ , \ \forall n\} \, , \tag{5.83}$$

is an algebra. The smallest σ-algebra in $\mathbb{R}^\infty$ that contains $\mathcal{C}$ is called the *Borel* σ-algebra in $\mathbb{R}^\infty$. We call a sequence of probability distributions $P_n(B_n)$ in $\mathbb{R}^n$ "compatible" if each $P_n(B_n)$ can be seen as a marginal distribution of P_{n+1}, that is, if

$$P_{n+1}\{\boldsymbol{\xi}^n \in B_n \ , \ \forall \xi_{n+1}\} \equiv P_n(\boldsymbol{\xi}^n \in B_n) \, . \tag{5.84}$$

The fundamental theorem of Kolmogorov takes the form:

Theorem 5.1 *There exists a unique probability distribution $P(C), C \in \mathcal{C}$ in $\mathbb{R}^\infty$ for which the compatible sequence $\{P_n(B_n) \ , \ B_n = P^n C\}$ has the function of the sequence of marginal distributions, that is, for each cylinder C_n defined by (5.82)*

$$P(C_n) \equiv P_n(B_n) . \tag{5.85}$$

It can be observed that, since $\{C_n\}$ is a sequence decreasing to C and $P_n(B_n)$ is a decreasing numerical sequence, it is clear that (5.85) implies

$$P(C) = \lim_{n \to +\infty} P_n(B_n) . \tag{5.86}$$

We have therefore constructed the probability space $(\mathbb{R}^\infty, \mathcal{C}, P)$ on which to base our calculations. In particular, with the definition of P we can proceed to the definition of mean and covariance of the r.v. Ξ, which as usual are given by

$$\mu = E\{\Xi\} = \int \xi d\, P(\xi) \tag{5.87}$$

$$C_\xi = E\{(\Xi - \mu)(\Xi - \mu)^\top\} . \tag{5.88}$$

Let's give at least one example of the construction of a probability distribution in $\mathbb{R}^\infty$.

Example 5.4 Let $\mu \in \mathbb{R}^\infty$ and we set $\mu_n = P^n \mu \in \mathbb{R}^n$; let $\{K_n\}$ be a sequence of covariance matrices, i.e., symmetric and positive definite, for which K_n is the principal minor of K_{n+1}. It is clear that when an element $k_{j\ell}$ enters K_n, it then remains constant in all $K_{n+1}, K_{n+2} \ldots$. We can therefore say that there exists an infinite-dimensional matrix K, i.e., an operator in $\mathbb{R}^\infty$, for which

$$K = \lim K_n . \tag{5.89}$$

We then consider the normal distributions in $\mathbb{R}^n$, $\forall n$

$$p_n(\xi) \sim \mathcal{N}(\mu_n, K_n) . \tag{5.90}$$

It is clear that the $p_n(\xi)$ satisfy the conditions of Kolmogorov's theorem (see (A.63)) and therefore there exists a probability space in $\mathbb{R}^\infty$ which we will say follows a normal distribution with mean μ and with covariance operator K,

$$P \sim \mathcal{N}(\mu, K) . \tag{5.91}$$

Note that while marginal distributions on finite subspaces can be defined by probability densities, this is not the case for the limit distribution in $\mathbb{R}^\infty$ as already mentioned.

Since within the scope of this text we will move exclusively in the linear field, for our purpose it is useful to define only a linear subspace of all the functions of a r.v. Ξ distributed over $\mathbb{R}^\infty$.

Definition 5.2 Let Ξ be a r.v. in $\mathbb{R}^\infty$ that admits mean μ and covariance K; we define the space

$$L^2(\Xi) \equiv \{U = a + \boldsymbol{a}^\top \Xi \; ; \; \forall a \in \mathbb{R} \, , \; \forall \boldsymbol{a} \in \mathbb{R}^\infty \; ; \; \boldsymbol{a}^\top K \boldsymbol{a} < +\infty\} \, . \quad (5.92)$$

We immediately note that if U is a r.v. in $L^2(\Xi)$, then U has finite mean and variance, i.e.,

$$\mu(U) = a + \boldsymbol{a}^\top \boldsymbol{\mu} \quad (5.93)$$

$$\sigma_U^2 = \boldsymbol{a}^\top K \boldsymbol{a} \; ; \quad (5.94)$$

therefore, using the Tchebychev theorem, applied to the variable Ξ, it is clear that

$$P(|U| = |a + \boldsymbol{a}^\top \Xi| < +\infty) = 1 \, . \quad (5.95)$$

Now we are interested in finding a sufficient condition so that we can affirm for a distribution P on $\mathbb{R}^\infty$, that for samples $\boldsymbol{\xi}$ drawn from this, $P(\boldsymbol{\xi} \in \ell^2) = 1$ holds.

A simple solution to this question is given by the following theorem.

Theorem 5.3 *Let P be the distribution of a r.v. Ξ in $\mathbb{R}^\infty$ with mean $\boldsymbol{\mu}$ and covariance K; if it results*

$$|\boldsymbol{\mu}|^2 + Tr K < +\infty \quad (5.96)$$

then for Ξ it holds

$$P(\Xi \in \ell^2) = 1 \, . \quad (5.97)$$

In fact, we have

$$\begin{aligned} E\{|\Xi|^2\} &= |\boldsymbol{\mu}|^2 + E\{|\Xi - \boldsymbol{\mu}|^2\} = \\ &= |\boldsymbol{\mu}|^2 + Tr E\{(\Xi - \boldsymbol{\mu})(\Xi - \boldsymbol{\mu})^\top\} = \\ &= |\boldsymbol{\mu}|^2 + Tr K \end{aligned} \quad (5.98)$$

and therefore by the Tchebychev theorem

$$P\{|\Xi|^2 > N^2\} \le \frac{E\{|\Xi|^2\}}{N^2} \, , \quad (5.99)$$

so that

$$0 \le P(\Xi \notin \ell^2) = P\{|\Xi|^2 > N^2 \, , \; \forall N\} = 0 \quad (5.100)$$

Equation (5.100) proves (5.97).

We observe that the condition $\boldsymbol{\xi} \in \ell^2$, almost certainly ($P = 1$), is equivalent to that $u \in H$, so in this way we have found a condition for P to be transported into a distribution on H, without resorting to the use of a space as large as that of all possible functions on T.

Now we can tackle the problem of explaining the meaning of an observation equation

$$X = \boldsymbol{m}^\top \boldsymbol{\xi} + \nu \, , \tag{5.101}$$

finding conditions for which the observable $\boldsymbol{m}^\top \boldsymbol{\xi}$ has a finite value almost certainly when $\boldsymbol{\xi}$ is a sample of Ξ whose probability distribution in $\mathbb{R}^\infty$ satisfies the condition (5.96).

Here too, instead of seeking the most general answer to the question posed, we are content with a sufficient condition that is easy to verify.

Lemma 5.4 *Let Ξ be a r.v. with distribution P for which the (5.96) holds; suppose that $\boldsymbol{\mu} \in \ell^2$ and that K is a bounded operator from ℓ^2 to ℓ^2, i.e.,*

$$\boldsymbol{m}^\top K \boldsymbol{m} \subset c\, \boldsymbol{m}^\top \boldsymbol{m} \, ; \tag{5.102}$$

then

$$\boldsymbol{m} \in \ell^2 \Rightarrow \boldsymbol{m}^\top \Xi \in L^2(\Xi) \, ; \tag{5.103}$$

therefore if Ξ and ν are independent and ν has finite variance σ_ν^2, then

$$P(|X| < +\infty) = 1 \, . \tag{5.104}$$

Indeed, if $|\boldsymbol{\mu}|^2 < +\infty$ and (5.102) holds, then (5.96) also holds and $\Xi \in \ell^2$ with $P = 1$.

Therefore, if $\boldsymbol{m} \in \ell^2$

$$E\{(\boldsymbol{m}^\top \Xi)^2\} = (\boldsymbol{m}^\top \boldsymbol{\mu})^2 + \boldsymbol{m}^\top K \boldsymbol{m} \le (\boldsymbol{m}^\top \boldsymbol{m})(\boldsymbol{\mu}^\top \boldsymbol{\mu}) + c\, \boldsymbol{m}^\top \boldsymbol{m} < +\infty \, . \tag{5.105}$$

Finally, assuming that Ξ and $\boldsymbol{\nu}$ are stochastically independent

$$E\{X^2\} = E\{(\boldsymbol{m}^\top \Xi)^2\} + \sigma_\nu^2 < +\infty \, , \tag{5.106}$$

so as usual, using the Tchebychev theorem, (5.104) holds.

What has been done for a single observation equation can be replicated for m different equations, giving meaning to the formula

$$\boldsymbol{X} = M\boldsymbol{\xi} + \boldsymbol{\nu} \tag{5.107}$$

$$(\boldsymbol{X}, \boldsymbol{\nu} \in \mathbb{R}^m \, , \; \boldsymbol{\xi} \in \mathbb{R}^\infty \, , \; M \cong (m, \infty))$$

with $\boldsymbol{\xi}$ a sample from Ξ.

Clearly, (5.107) makes sense, that is

$$P(|\boldsymbol{X}| < +\infty) = 1\,, \tag{5.108}$$

if (sufficient condition) the m rows of M,

$$M = \begin{vmatrix} \boldsymbol{m}_1^\top \\ \vdots \\ \boldsymbol{m}_m^\top \end{vmatrix}\,, \tag{5.109}$$

are all in ℓ^2, or in other words, M is a bounded operator $\ell^2 \to \mathbb{R}^m$. At this point we have the tools to look for a Bayesian "estimator", $\widehat{\boldsymbol{\xi}}(\boldsymbol{X})$, of Ξ, based on the observation equation (5.12). We note that we do not address here the more complex issue of the conditional distribution of Ξ given $\boldsymbol{X}$, which would force us to discuss the prior of Ξ on $\mathbb{R}^\infty$. When one would want to use a non-informative prior, one would indeed run into the problem of the non-existence of a uniform distribution on $\mathbb{R}^\infty$, whose solution would then lead to the introduction of the concept of generalized random field, which is beyond the scope of this book.

Here we limit ourselves to a theory of *linear regression*, in the Bayesian sense, of Ξ on X, well known in geophysical and geodetic literature (see [11, 43, 46, 52]).

So we look for a Bayesian "estimator" of Ξ in the form

$$\widehat{\boldsymbol{\xi}}(\boldsymbol{X}) = \Lambda \boldsymbol{X} \tag{5.110}$$

with $\boldsymbol{X}$ given by (5.12). The idea is to use Λ to minimize the total error of the estimator $\widehat{\boldsymbol{\xi}}$,

$$\mathcal{E}_t^2 = E\{|\Xi - \widehat{\boldsymbol{\xi}}|^2\}\,. \tag{5.111}$$

For simplicity, we assume that

$$E\{\Xi\} = \boldsymbol{\mu} = 0\,, \tag{5.112}$$

otherwise it would be enough to replace $\boldsymbol{X} - M\boldsymbol{\mu}$ with $\boldsymbol{X}$ in (5.12). This assumption also implies that $E\{\boldsymbol{X}\} = 0$.

We also assume that $TrK < +\infty$, implying that K satisfies (5.102) too.

Now we calculate (5.111) as

$$\mathcal{E}_t^2 = TrK - 2Tr\Lambda C_{X,\Xi} + Tr\Lambda C_X \Lambda^\top\,, \tag{5.113}$$

where, from (5.110)

$$C_{X,\Xi} = E\{\boldsymbol{X}\Xi^\top\} = ME\{\Xi\Xi^\top\} = MK \tag{5.114}$$

and

$$C_X = MKM^\top + C_\nu \,. \tag{5.115}$$

The minimization of (5.113) with respect to Λ leads to the usual linear regression equation (for variables with zero mean)

$$\Lambda = C_{\Xi X} C_X^{-1} = KM^\top (MKM^\top + C_\nu)^{-1} \,, \tag{5.116}$$

which, when substituted into (5.113), gives for $\mathcal{E}_t^2$ the expression

$$\mathcal{E}_t^2 = TrK - TrKM^\top (MKM^\top + C_\nu)^{-1} MK \,. \tag{5.117}$$

So the Bayesian "estimator" $\widehat{\boldsymbol{\xi}}$ according to (5.110) is given by

$$\widehat{\xi} = KM^\top (MKM^\top + C_\nu)^{-1} \boldsymbol{X} \,. \tag{5.118}$$

A comparison with the regularized frequentist solution (5.66) allows us to prove the following Lemma of equivalence.

Lemma 5.5 *The Bayesian solution (5.118) is formally equivalent to the regularized frequentist solution (5.66), provided that*

$$\lambda Q = K^{-1} \,. \tag{5.119}$$

In fact, it is sufficient to verify algebraically that

$$KM^\top (MKM^\top + C_\nu)^{-1} = (M^\top C_\nu^{-1} M + K^{-1})^{-1} M^\top C_\nu^{-1} \,, \tag{5.120}$$

multiplying on the left by $(M^\top C_\nu^{-1} M + K^{-1})$ and on the right by $(MKM^\top + C_\nu)$.

Note that the equivalence is formal in the sense that in (5.66) $\boldsymbol{X}$ is a r.v. with non-zero mean that is stochastic only due to the presence of $\boldsymbol{\nu}$, while in (5.118) $\boldsymbol{X}$ is a r.v. with zero mean that depends on the two stochastic variables Ξ and $\boldsymbol{\nu}$.

However, as we can see, the Bayesian approach eliminates the problem of determining the optimal value of λ a posteriori.

Example 5.5 Let's revisit Examples 5.2 and 5.3 of determining a global model of anomalous gravitational potential. First, we need to define $\boldsymbol{\mu}$ and K for the variable Ξ which here coincides with the vector $T_{j\ell}$.

Suppose that

$$E\{T_{j\ell}\} = 0 \tag{5.121}$$

$$E\{T_{j\ell} T_{rs}\} = \sigma_j^2 \delta_{jr} \delta_{\ell s} \,; \tag{5.122}$$

hence the $\{T_{j\ell}\}$ are a priori linearly independent and their variance depends only from the degree j and not from the order m. It is possible to demonstrate that this a priori assumption is linked to a hypothesis of stochastic invariance of $\{T(r, \sigma)\}$ under the group of rotations (see [43, 47]).

The a priori values σ_j^2 can be derived with an interpolation model of the empirical values of the so-called degree variances (see [47]). Using the equivalence Lemma and the (5.79) it is seen that the Bayesian solution in this case is given by

$$\widehat{T}_{j\ell} = \left(\frac{\mu}{R^2}\right)^{-1} \frac{j-1}{(j-1)^2 + 1/\sigma_j^2} \Delta g_{oj\ell} \,. \tag{5.123}$$

Thus we have that, since $\sigma_j^2 \to 0$, the (5.123) tends to strongly regularize $T_{j\ell}$ for high values of the degree j, compared to a simple ridge regression.

Chapter 6
A Look at Machine Learning

Machine learning is defined as the algorithmic structure of the wider area of Artificial Intelligence. The fact that machine learning is actually an extremization of black box theory is highlighted and the consequent attitude towards an over-parametrization of models is presented with the tools necessary to solve the corresponding optimization problem. Deep machine learning, with its relation to neural networks and logistic regression is then examined. The main theorem of approximation of continuous or integrable functions by a finite chain of functions is established, some examples illustrate possible applications of the above concept to Earth Sciences.

6.1 Introduction and Problem Statement

We have already said in Chap. 2 that in the probabilistic view of the world, i.e., of the stochastic models that allow predicting the result of experiments beyond those already carried out, there is a spectrum of interpretations of the stochastic model depending on whether it is believed to describe a process that follows known physical laws, but containing so many variables as to be impossible to control [17] or that the model is so to speak created by the large amount of available data, just listening to "*what data says*" [19].

The "machine learning" is conceptually an evolution of this second attitude and ultimately of the theory of systems and the idea of "black box".

From Wikipedia (2024), we read:

> In science, computing and engineering, a black box is a system which can be viewed in terms of its inputs and its outputs (or transfer characteristics) without knowledge of its internal working. Its implementation is opaque.

F. Sansò, A. Albertella, *The Probabilistic Vision of the Physical World*, Lecture Notes in Geosystems Mathematics and Computing,
https://doi.org/10.1007/978-3-031-88268-5_6

The use of this concept and this approach has enormously grown with the increase in memory and computing capabilities, as well as the possibility of acquiring and therefore managing huge amounts of data almost in every field, leading to the research and application sector known as Artificial Intelligence. Quoting, perhaps not by chance, an economist, Will Kenton [https://www.investopedia.com/terms/b/blackbox.asp]

> [...] Advances in computing power, artificial intelligence and machine learning capabilities are causing a proliferation of black boxes models in many professions and are adding to the mystic surrounding them.

And again from Maurizio di Paolo Emilio [https://www.innovationpost.it/tecnologie/intelligenza-artificiale-deep-learning-e-machine-learning-quali-sono-le-differenze]:

> Artificial intelligence involves all those operations characteristic of human intellect performed by computers; including understanding language, recognizing objects and sounds, learning and problem solving. [...] "Machine learning" is a path to the implementation of Artificial Intelligence. "Deep machine learning" is one of the approaches to automatic learning, which was inspired by the structure of the brain, i.e., from the interconnection of various neurons.

So "machine learning" and "deep machine learning", which is a specification of it, are the algorithmic part of the application of artificial intelligence, which however is a much broader field that includes the choice of variables necessary to describe a system and the rules of numerical coding of these variables, necessary for their automatic processing, as well as the choice of objective functions to optimize and finally the interpretation of the results, i.e., the decoding from numbers to variables that can also have a purely qualitative meaning.

In our opinion, these choices make the processes of Artificial Intelligence, a name that we find inappropriate, have fueled in the public that mystical aura that has led to the discussion that we believe is misleading, on the autonomy of computer intelligence, borrowed more from science fiction than from science.

In any case, having our gaze turned to disciplines of a physical nature such as Earth sciences, we here focus on the algorithmic part, trying to grasp its logical structure and its contribution to the probabilistic view of the world. The following definitions are borrowed from the remarkable book by Vapnik [56]. We immediately say that conceptually the approach described here is in a certain sense a specification of what was presented in Chap. 3. First of all, the observables are now divided into two parts $(\boldsymbol{Y}, \boldsymbol{X})$ and it is assumed that together they form a vector r.v., with its probability distribution, $P(\boldsymbol{Y}, \boldsymbol{X})$, which however is not known. The difference between the two components $\boldsymbol{Y}$ and $\boldsymbol{X}$ is that the vector $\boldsymbol{X}$ is considered as an input variable of a process, while $\boldsymbol{Y}$ is an output variable, i.e., $\boldsymbol{Y}$ is thought of as a variable dependent, albeit stochastically, on $\boldsymbol{X}$. For simplicity of exposition here we will always consider the case where Y is a one-dimensional variable.

The data we have are those of a Bernoulli sample $(y_i, \boldsymbol{x}_i)$, with $i = 1, \dots N$, drawn from $(Y, \boldsymbol{X})$. The problem we want to solve can take two forms:

a) the direct problem similar to those already described in Chap. 3, consists in finding a function of X

$$\widehat{Y} = g(\boldsymbol{X}) \tag{6.1}$$

that explains as much as possible Y. This is ultimately a regression problem, which we can also write as

$$Y = \widehat{Y} + \varepsilon \equiv g(\boldsymbol{X}) + \varepsilon \ , \tag{6.2}$$

with the idea that the discrepancy ε should be a "small" r.v. To give meaning to the adjective "small" we can observe that

$$\varepsilon = \varepsilon(Y, \boldsymbol{X}) = Y - g(\boldsymbol{X}) \tag{6.3}$$

and define a "loss function" $\mathcal{L}$,

$$\mathcal{L}(Y, \widehat{Y}) = \mathcal{L}(\varepsilon) = \mathcal{L}(Y - g(\boldsymbol{X})) \ ; \tag{6.4}$$

subsequently we define a theoretical risk function $\mathcal{R}$,

$$\mathcal{R}(g) = E_P\{\mathcal{L}(Y - g(\boldsymbol{X}))\} \tag{6.5}$$

which leads to considering a risk, corresponding to the g that makes (6.5) minimum, as conveniently small.
In (6.5) the index P means that the average is taken with respect to the distribution $P(Y, \boldsymbol{X})$; naturally this is unknown and therefore we used the term "theoretical" for the risk $\mathcal{R}(g)$.
The difference with the discussion of Chap. 3 is that here now $\boldsymbol{X}$ is not a vector of parameters but a vectorial r.v. of which we possess the sample $(\boldsymbol{x}_i)$,

b) the indirect problem consists in the search for an estimate $Q(Y, \boldsymbol{X})$ of the distribution $P(Y, \boldsymbol{X})$, which subsequently can allow to construct, in an approximate way, the distribution $P(Y|\boldsymbol{X})$.

This in turn could be used to solve the direct problem, **a)**, but in particular it is very useful in classification problems, in which the dependent variable Y is discrete

$$Y \equiv \{k \ ; \ k = 1 \dots K\} \ ; \tag{6.6}$$

in the distribution $P(k, \boldsymbol{x})$ the argumental value $(k, \boldsymbol{x})$ indicates that the element $\boldsymbol{x}$ belongs to the class k. The Neyman-Pearson lemma indeed allows to determine the optimal estimator of k by comparing the $P(Y = k|\boldsymbol{X} = \boldsymbol{x})$ among themselves.

To solve problem **b)** we need to introduce a notion of distance between the estimator $Q(Y, \boldsymbol{X})$ and the true distribution; in literature, [3], we can have various forms among which maybe the easiest is the cross-entropy

$$H(P, Q) = -E_P\{\log Q\} = -\int P(y, \boldsymbol{x}) \log Q(y, \boldsymbol{x}) \, dy \, d\boldsymbol{x} \tag{6.7}$$

with the warning that H is not a true distance, because $H(P, Q) \neq H(Q, P)$, although it is shown that it becomes minimum for $Q \equiv P$.

6.2 Machine Learning

We premise that in this section we tackle the solution of the direct problem and the indirect one using a parametric approach. Therefore the regression function for problem **a)** will have the form

$$\widehat{Y} = g(\boldsymbol{X}|\boldsymbol{\vartheta}) , \tag{6.8}$$

while the distribution $Q(Y, \boldsymbol{X})$ that we will use to approximate the $P(Y, \boldsymbol{X})$, problem **b)**, will have the form

$$Q = Q(Y, \boldsymbol{X}|\boldsymbol{\vartheta}) . \tag{6.9}$$

We note that the idea of minimizing the (6.5) for the direct problem and the (6.7) for the indirect problem, is not feasible in this form, as the functional $E_P\{\ \}$ depends on P, which in our case is unknown. However, using the CLT (see Theorem A.5), $E_P\{\ \}$ can be replaced by a sample average, especially when the sample is very large. Therefore we introduce an empirical risk

$$\mathcal{R}_e(\boldsymbol{\vartheta}) = \frac{1}{N} \sum_{i=1}^{N} \mathcal{L}(Y_i, g(\boldsymbol{X}_i|\boldsymbol{\vartheta})) \tag{6.10}$$

and an empirical cross-entropy

$$H_e(\boldsymbol{\vartheta}) = -\frac{1}{N} \sum_{i=1}^{N} \log Q(Y_i, \boldsymbol{X}_i|\boldsymbol{\vartheta}) \tag{6.11}$$

and we will indicate as solution of the two problems **a)** and **b)** the minimization respectively of $\mathcal{R}_e(\boldsymbol{\vartheta})$ and of $H_e(\boldsymbol{\vartheta})$. Now it should be noted that in general we cannot assume that the $g(\boldsymbol{X})$ of the relation (6.1) belongs to the family $\{g(\boldsymbol{X}|\boldsymbol{\vartheta})\}$, for some $\boldsymbol{\vartheta}$; similarly we cannot expect that $P(Y, \boldsymbol{X})$ belongs to the family $Q(Y, \boldsymbol{X}|\boldsymbol{\vartheta})$. Therefore in the estimation procedure there will remain a bias, both in the minimization of (6.10) and in that of (6.11). Naturally, not being able to know a priori how large the bias is, one is driven to choose families with many parameters, ideally so that, if the number of parameters tends to infinity, the bias tends to zero. This argument will be resumed in the next section. Here we are interested in observing that if the minimum problem becomes unstable or even underdetermined due to the presence of too many parameters, then it is natural to modify the functional to be minimized by adding a regularization term, $\lambda|\boldsymbol{\vartheta}|^2$, as already done in Chap. 5, see (5.58). Therefore the optimization principles of problems **a)** and **b)** become the search for the minimum of

$$J_\lambda(\boldsymbol{\vartheta}) = \mathcal{R}_e(\boldsymbol{\vartheta}) + \lambda|\boldsymbol{\vartheta}|^2 \tag{6.12}$$

for the direct problem (regression), and

$$K_\lambda(\boldsymbol{\vartheta}) = H_e(\boldsymbol{\vartheta}) + \lambda|\boldsymbol{\vartheta}|^2 \tag{6.13}$$

for the indirect problem.

This leaves open the problem of determining λ. The preferred solution in this case is not of Morozov's functional type, mentioned in Chap. 5, see (5.58), but rather follows a completely different approach that provides more guarantees from a statistical point of view, as it uses different data from those used to obtain the estimate $\widehat{\boldsymbol{\vartheta}}$. The point is to divide from the beginning the available sample of N elements, $(y_i, \boldsymbol{x}_i)$, into two parts that we call training and test of size N_{Tr} and N_{Te}, so that

$$N = N_{Tr} + N_{Te}\ ; \tag{6.14}$$

the values corresponding to $i = 1 \dots N_{Tr}$ identify the training sample, those with $i = N_{Tr} + 1 \dots N$ identify the test sample.

The training sample is used to minimize $J_\lambda(\boldsymbol{\vartheta})$ or $K_\lambda(\boldsymbol{\vartheta})$, for each fixed value of λ. This creates, in principle, the function

$$\boldsymbol{\vartheta} = \boldsymbol{\vartheta}(\lambda)\ . \tag{6.15}$$

Now we define the cross-variance

$$CV(\boldsymbol{\vartheta}) = \frac{1}{N_{Te}} \sum_{N_{Tr}+1}^{N} \mathcal{L}(Y_i, g(\boldsymbol{X}_i|\boldsymbol{\vartheta})) \tag{6.16}$$

for problem **a**), or the cross-log-likelihood

$$C\ell(\boldsymbol{\vartheta}) = -\frac{1}{N_{Te}} \sum_{N_{Tr}+1}^{N} \log Q(Y_i, \boldsymbol{X}_i|\boldsymbol{\vartheta}) \tag{6.17}$$

for problem **b**). As can be seen, both of these functions are empirical, i.e., based on sample values.

Substituting (6.15) into (6.16) we create

$$CV(\lambda) = CV[\vartheta(\lambda)] \tag{6.18}$$

while with (6.17) we have

$$C\ell(\lambda) = C\ell[\vartheta(\lambda)]\ . \tag{6.19}$$

The optimal value of λ for problem **a)** or for problem **b)** is, respectively, the one that minimizes $CV(\lambda)$ or $C\ell(\lambda)$. As for the proportion of the original sample to use for training or testing, the most common suggestion in the literature (see for example [3]) is that the test sample should have a significant size compared to the total sample, up to 50%. It can be added here that the sample values of the $\{\boldsymbol{x}_i\}$ for testing should be as mixed as possible with those for training.

As can be seen, the solution to the machine learning problem, in forms **a)** and **b)**, is reduced to finding the minimum of a function of $\boldsymbol{\vartheta}$. This classic optimization problem can become very burdensome to solve numerically, especially if the number of parameters $\boldsymbol{\vartheta}$ is high and if the objective function has many relative minima. In the latter case, one must resort to methods that go beyond the objectives of this presentation (see, in this regard, [51]). In any case, especially for large $\boldsymbol{\vartheta}$ and for large values of N, it is important to use a method that is based only on the calculation of the first derivatives of the function to be minimized, avoiding the calculation of the second derivatives, as required for example by the Newton–Raphson method.

In this respect, we briefly analyze in Appendix C some methods used for this purpose. We report here the conclusions obtained from the discussion presented in that appendix. For convenience of notation we put

$$F(\boldsymbol{\vartheta}, \lambda) = \frac{1}{N_{Tr}} \sum_{i=1}^{N_{Tr}} f_i(\boldsymbol{\vartheta}) + \lambda |\boldsymbol{\vartheta}|^2 \tag{6.20}$$

with

$$f_i(\boldsymbol{\vartheta}) = f(Y_i, \boldsymbol{X}_i|\boldsymbol{\vartheta}) = \begin{cases} [Y_i - g(\boldsymbol{X}_i|\boldsymbol{\vartheta})]^2 \\ \text{(direct problem)} \\ -\log Q(Y_i, \boldsymbol{X}_i|\boldsymbol{\vartheta}) \\ \text{(indirect problem)} \end{cases} \tag{6.21}$$

For a fixed λ, the minimum of $F(\boldsymbol{\vartheta}, \lambda)$ with respect to $\boldsymbol{\vartheta}$ can be obtained as the limit of the sequence $\{\boldsymbol{\vartheta}_n\}$ defined by the steepest descent algorithm

$$\boldsymbol{\vartheta}_{n+1} = \boldsymbol{\vartheta}_n - \gamma_n \nabla_{\boldsymbol{\vartheta}} F(\boldsymbol{\vartheta}_n, \lambda) \,. \tag{6.22}$$

In (6.22) the sequence $\{\gamma_n\}$ must satisfy the conditions

$$\sum_{n=0}^{+\infty} \gamma_n = +\infty \,, \quad \sum_{n=0}^{+\infty} \gamma_n^2 < +\infty \,, \tag{6.23}$$

for example by setting

$$\gamma_n = \frac{c}{n+1} \,. \tag{6.24}$$

The implementation of (6.22) requires the calculation of $\nabla_{\vartheta} f_i(\boldsymbol{\vartheta})$ a number of times equal to N_{Tr}, which can easily be of the order of 10^6. However, since $F(\boldsymbol{\vartheta}, \lambda)$ has the form of an empirical average on the training sample, a good approximation of $\nabla_{\vartheta} F(\boldsymbol{\vartheta}, \lambda)$ can be achieved by randomly extracting a mini-batch, of size M equal to a few hundred, and calculating the so-called stochastic gradient

$$\nabla_{\vartheta} \widetilde{F}(\boldsymbol{\vartheta}, \lambda) = \frac{1}{M} \sum_{i=1}^{M} \nabla_{\vartheta} f_i(\boldsymbol{\vartheta}) + 2\lambda \boldsymbol{\vartheta} , \tag{6.25}$$

which involves a more feasible numerical effort.

We also note that the problem of the two concatenated minima of the functions (6.16)–(6.18) and (6.17)–(6.19) presents some additional difficulty, therefore in Appendix D a possible iterative solution of the two problems is reported.

In conclusion, we arrive at the following statement of equivalence:

The direct problem **a)** of machine learning and the indirect problem **b)** are equivalent respectively to the problem of regression with the method of regularized least squares and to the problem of maximum likelihood, in this case regularized, seen in Chap. 3, as far as the estimation of the parameters $\boldsymbol{\vartheta}$ is concerned; the estimation of the regularization parameter λ is carried out with a second optimization coupled to that for $\boldsymbol{\vartheta}$ (see Appendix D).

6.3 Deep Machine Learning

Deep machine learning consists in the construction of a model, in our case parametric, of machine learning, in which the estimation of the parameters is done by means of a "neural network" (NN), which implements the optimization of the target function. In turn, a neural network is an algorithm for the construction of a non-linear model of function of the parameters, whether it is a $f(\boldsymbol{X}|\boldsymbol{\vartheta})$ in the case of regression or $Q(Y, \boldsymbol{X}|\boldsymbol{\vartheta})$ in the case of approximation of a probability density, combining different functions of functions.

Example 6.1 For example, suppose we want to find the regression function

$$y = f(\boldsymbol{x}|\boldsymbol{\vartheta}) , \tag{6.26}$$

defining the family $f(\boldsymbol{x}|\boldsymbol{\vartheta})$ as a combination of three functions, generally non-linear

$$f(\boldsymbol{x}|\boldsymbol{\vartheta}) = f_3(f_2(f_1(\boldsymbol{x}|\boldsymbol{\alpha})|\boldsymbol{\beta})|\boldsymbol{\gamma}) . \tag{6.27}$$

This can be interpreted as a chain of non-linear transformations that, starting from $\boldsymbol{x}$, define new variables, called *hidden variables*,

$$\boldsymbol{h} = f_1(\boldsymbol{x}|\boldsymbol{\alpha}) , \; \boldsymbol{z} = f_2(\boldsymbol{h}|\boldsymbol{\beta}) , \tag{6.28}$$

arriving at the output variable y

$$y = f_3(z|\gamma)\,. \tag{6.29}$$

In this case the global vector of parameters is given by

$$\boldsymbol{\vartheta}^\top = (\boldsymbol{\alpha}^\top, \boldsymbol{\beta}^\top, \boldsymbol{\gamma}^\top)^\top\,. \tag{6.30}$$

The name neural network (NN) comes from the analogy with the nervous networks of the human body and the transmission through this of sensory impulses (input) to the neurons of the brain, until creating a reaction (output).

In particular, NNs lend themselves well to a graphical representation that is very expressive. The NN is represented by a graph divided by levels: the input level, i.e. the vector $\boldsymbol{X}$, the hidden levels, i.e. the variables $\boldsymbol{h}$ and $\boldsymbol{z}$ in Example 6.1, the output level, i.e. the variable Y.

The units of a level are connected to those of the lower level and the upper level, although, in a more advanced form, there can be horizontal connections or those that skip levels, which we will not consider here. The connections between a level and the upper level are usually realized by a linear transformation combined with a non-linear function, called *activation function*. Typically the activation function has the same form for all units of the level. We explicitly underline that in this section we use the symbol h to define the units of a hidden level (hidden layer), not to be confused with the symbol $\boldsymbol{h} = \nabla_{\boldsymbol{\vartheta}} f(\boldsymbol{X}|\boldsymbol{\vartheta})$ used in Appendices C and D.

We present in Example 6.2 a graph related to Example 6.1, which we believe to be self-explanatory.

Example 6.2 We take up Example 6.1, in Fig. 6.1

$$h_1 = f_1(\sum_{i=1}^{3} \alpha_{i1} X_i)$$

$$h_2 = f_1(\sum_{i=1}^{3} \alpha_{i2} X_i)$$

$$z_1 = f_2(\sum_{i=1}^{2} \beta_{i1} h_i)$$

$$z_2 = f_2(\sum_{i=1}^{2} \beta_{i2} h_i)$$

$$y = f_3(\sum_{i=1}^{2} \gamma_i Z_i)$$

f_1, f_2, f_3 are activation functions.

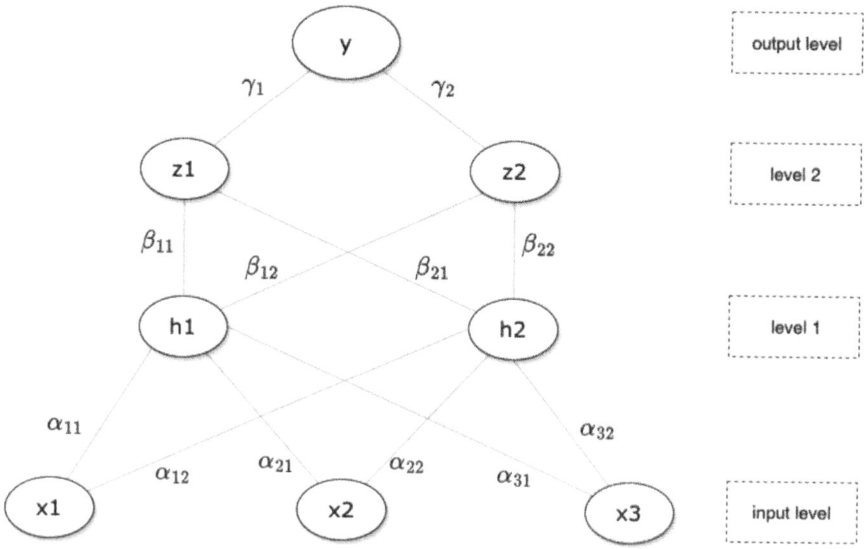

Fig. 6.1 Graph of a NN. The circles indicate the units and the lines indicate the connections

As seen in Example 6.2, the NN can be fully connected, which brings to 12 the number of parameters that define its model. Of course, it is not said that all connections must be active and the parameters could be constrained for reasons of symmetry of dependencies. For example, we could have

$$\alpha_{11} = \alpha_{21} = \alpha_{22} = \alpha_{32}\ ,\ \ \alpha_{12} = \alpha_{31}$$
$$\beta_{11} = \beta_{22}\ ,\ \ \beta_{12} = \beta_{21}$$
$$\gamma_1 = \gamma_2\ .$$

This leads to a reduction of the system parameters. Sometimes it is desired that the linear transformations between one level and the next are non-homogeneous, i.e., they contain an additive constant. This is equivalent to introducing, for example at the input level, a variable X_0 that always takes the value 1. The same can be done for the other hidden levels, so formally we always return to a homogeneous linear model in the parameters, extending the summations from 0 instead of from 1. In Fig. 6.2 we report some of the most commonly used activation functions in the NN literature. In particular, the sigmoid functions, i.e., such that

$$\lim_{t\to-\infty} \sigma(t) = 0\ ,\quad \lim_{t\to+\infty} \sigma(t) = 1\ , \tag{6.31}$$

like the one in Fig. 6.2a), are often used in NN applications.

One might wonder if models of the type (6.27) are general enough to approximate any continuous or just integrable regression function. The problem was discussed by

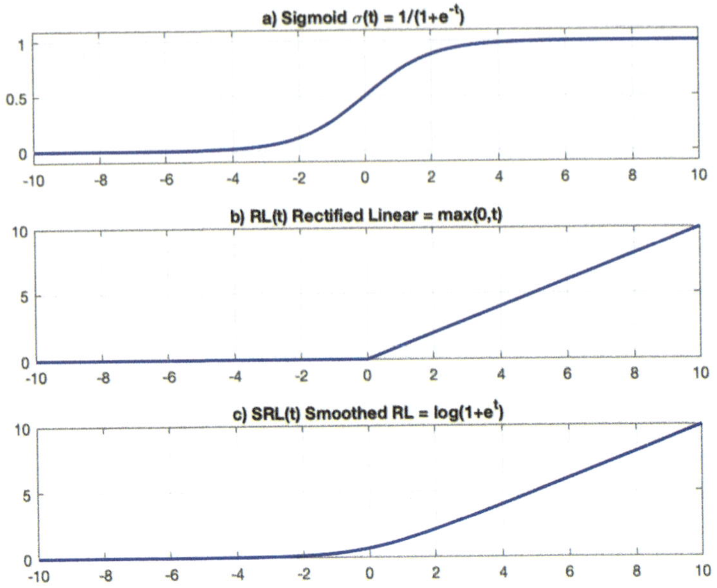

Fig. 6.2 Most commonly used activation functions. (**a**) Sigmod $\sigma(\mathrm{t}) = 1/(1+\mathrm{e}^{-\mathrm{t}})$. (**b**) RL(t) rectified linear = max(0,t). (**c**) SRL(t) smoothed RL = $\log(1+\mathrm{e}^{\mathrm{t}})$

A. N. Kolmogorov as early as 1957, and subsequently solved for sigmoid activation functions by G. Cybenko in 1989, affirmatively even for the simplest NN, the one with only one hidden level, the so-called vanilla NN (see [9]). We report without proof the Cybenko theorem, modifying its notation to bring it back to the one used here.

Theorem 6.1 *Let I_n be the closed unit cube in $\mathbb{R}^n$ and $C(I_n)$ the Banach space of continuous functions in I_n, with the topology of the maximum, i.e.,*

$$\parallel f(\boldsymbol{x}) \parallel_{C(I_n)} = \max_{x \in I_n} |f(\boldsymbol{x})| \,, \tag{6.32}$$

also let $\sigma(t)$ be a sigmoid activation function, i.e., satisfying (6.31), and continuous on $\mathbb{R}$, then the family

$$\begin{cases} g(\boldsymbol{x}|\{\alpha_{j\ell}\}\,,\ \{\gamma_\ell\}\,,\ N_h) = \sum_{\ell=0}^{N_h} \gamma_\ell \sigma(\sum_{j=0}^{n} \alpha_{j\ell} x_j) \\ \forall N_h \text{ integer, } \{\alpha_{j\ell}\} \in \mathbb{R}^N \otimes \mathbb{R}^n\,,\ \{\gamma_\ell\} \in \mathbb{R}^N \end{cases} \tag{6.33}$$

is dense in $C(I_n)$.

This means that every $f(\boldsymbol{x}) \in C(I_n)$ can be uniformly approximated as desired by a combination of the type (6.33) provided that $N_h, \{\alpha_{j\ell}\}, \{\gamma_\ell\}$ are chosen appropriately.

We observe that the combinations of functions of (6.33) are precisely those produced by a vanilla NN with N_h the number of units of the hidden level; see Fig. 6.1 imagining removing the hidden level of the z. Typically, a large N_h will need to be chosen, i.e., a wide width for the hidden level. However, the application of Theorem 6.1 to deep machine learning requires, in our opinion, a slightly more accurate analysis than that presented by Cybenko in [9]. In fact, particularly for regression, based on the theorem it can be stated that, if

$$y = f(\boldsymbol{x}) \tag{6.34}$$

is the function we are looking for and the sample data are therefore

$$Y_i = f(\boldsymbol{X}_i)\ i = 1 \ldots N \tag{6.35}$$

then the function

$$v(\boldsymbol{x}|\boldsymbol{\vartheta}) = y - g(\boldsymbol{x}|\boldsymbol{\vartheta}) = f(\boldsymbol{x}) - g(\boldsymbol{x}|\boldsymbol{\vartheta}) \tag{6.36}$$

can be reduced in absolute value to less than an arbitrary ε for an appropriate $\boldsymbol{\vartheta}$ and for all $\boldsymbol{x} \in I_n$. It follows that

$$\mathcal{R}_e(\boldsymbol{\vartheta}, \lambda) = \frac{1}{N}\sum_{i=1}^{N} v(\boldsymbol{x}_i|\boldsymbol{\vartheta})^2 + \lambda|\vartheta|^2 \leq \varepsilon^2 + \lambda|\boldsymbol{\vartheta}|^2 . \tag{6.37}$$

As can be seen, given an $\eta > 0$, the empirical risk

$$\mathcal{R}_e(\boldsymbol{\vartheta}, \lambda) < \eta \tag{6.38}$$

can only be reduced by choosing λ appropriately small. This is a distinctive feature of Tikhonov's approach to the regularization of ill-posed problems (see [54]).

We also note that the possible introduction of white noise, $\{\nu_i\}$, in Eq. (6.35), does not change our conclusions, as is easy to see.

Now two elementary examples to understand how NNs can be designed to solve problems related to Earth sciences.

Example 6.3 A rain gauge p is located in the center of an area where 3 weather stations are also located (Fig. 6.3). The stations measure (T, P, e), respectively air temperature, pressure and relative humidity. The collected data are average hourly values; we set

$Y_i = p_i$ rain level at time i

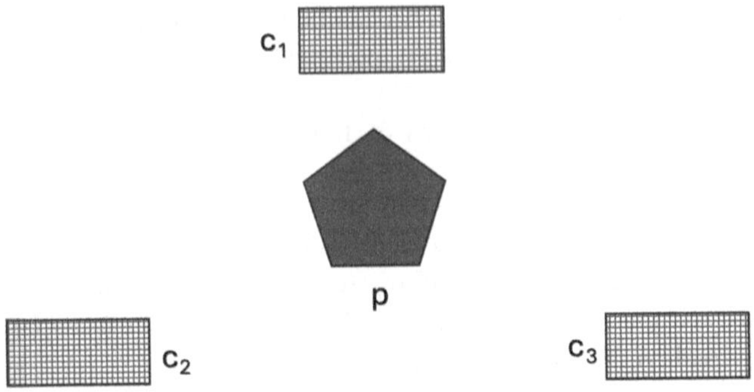

Fig. 6.3 The configuration of the rain gauge p and the stations c_1, c_2, c_3

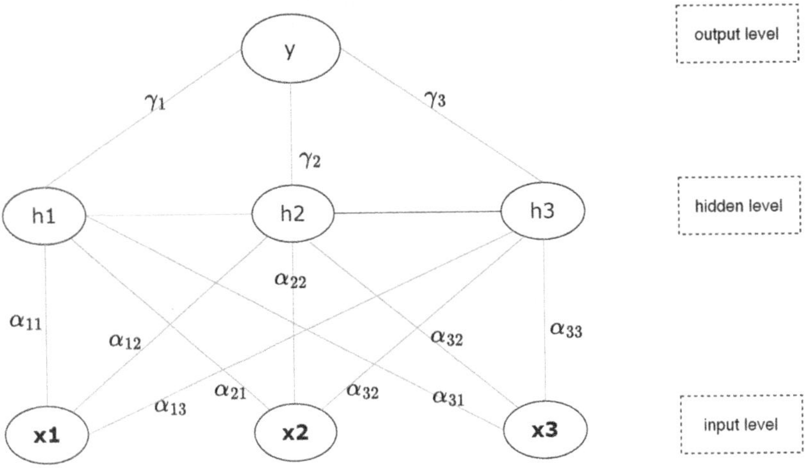

Fig. 6.4 The neural network representing the family $g(\boldsymbol{X}_1, \boldsymbol{X}_2, \boldsymbol{X}_3|\boldsymbol{\vartheta})$

$\boldsymbol{X}_{ki} = (T, P, e)_{ki}$ weather parameters (temperature, pressure, air humidity) of station c_k at time i

and we want to build a model for the regression of p on $(\boldsymbol{X}_1, \boldsymbol{X}_2, \boldsymbol{X}_3)$

$$p = f(\boldsymbol{X}_1, \boldsymbol{X}_2, \boldsymbol{X}_3) \tag{6.39}$$

based on a sample of size N.

The NN representing the family $g(\boldsymbol{X}_1, \boldsymbol{X}_2, \boldsymbol{X}_3|\boldsymbol{\vartheta})$ with which to approximate $f(\boldsymbol{X}_1, \boldsymbol{X}_2, \boldsymbol{X}_3)$ could be like the one shown in Fig. 6.4 (vanilla NN).

Note that the input units are vectorial while in the hidden units we have set scalar variables, so the link between level $\boldsymbol{X}$ and level h can be written as

$$h_\ell = \sigma\left(\sum_{k=1}^{3} \boldsymbol{\alpha}_{k\ell}^\top \boldsymbol{X}_k + \alpha_{0\ell}\right) , \quad \boldsymbol{\alpha}_{k\ell} \in \mathbb{R}^3 , \quad \alpha_{0\ell} \in \mathbb{R} ,$$

having chosen a sigmoid activation function, like the one in Fig. 6.2a).

The link between Y and h will then be

$$Y = \sum_{\ell=1}^{3} \gamma_\ell h_\ell + \gamma_0 .$$

These two concatenated relations must then be repeated for each index i that runs on the sample $(Y_i, \boldsymbol{X}_{1i}, \boldsymbol{X}_{2i}, \boldsymbol{X}_{3i})$. So the function

$$J_\lambda(\boldsymbol{\vartheta}) = J_\lambda(\gamma_0, \gamma_\ell, \alpha_{0\ell}, \boldsymbol{\alpha}_{k\ell}), \quad (\ell, k = 1, 2, 3) \tag{6.40}$$

to be optimized will be

$$J_\lambda(\boldsymbol{\vartheta}) = \frac{1}{N}\sum_{i=1}^{N}\{Y_i - [\sum_{\ell=1}^{3}\gamma_\ell\, \sigma(\sum_{k=1}^{3}\boldsymbol{\alpha}_{k\ell}^\top \boldsymbol{X}_{ki} + \alpha_{0\ell}) + \gamma_0]\}^2 +$$

$$+\lambda\{\gamma_0^2 + \sum_{\ell=1}^{3}\gamma_\ell^2 + \sum_{\ell=1}^{3}\alpha_{0\ell}^2 + \sum_{\ell=1}^{3}\sum_{k=1}^{3}|\boldsymbol{\alpha}_{k\ell}|^2\} .$$

We will see in the following of the section how to apply the steepest descent algorithm using the NN in Fig. 6.4.

Example 6.4 Consider the problem of the probabilistic model of classification into two classes, for example that of a black and white image. A problem of this kind occurs for example in forestry sciences, when a forest is divided into (regular) pixels and a pixel is blackened if the presence of a certain disease is found in the plants inside it. The aim is to build a parametric model $q(Y, \boldsymbol{X}|\boldsymbol{\vartheta})$ where

$$Y = \begin{cases} 1 & \text{black pixel} \\ 0 & \text{white pixel} \end{cases}$$

$$\boldsymbol{X} = \begin{vmatrix} X_1 \\ X_2 \end{vmatrix} \quad \text{position of the pixel under examination.}$$

The sample used will be $(Y_i, \boldsymbol{X}_i)$, $i = 1 \dots N$, where $\boldsymbol{X}_i$ is the position of the i-th pixel of the sample.

Note that, given the binary nature of Y, we can write

$$p(Y|\boldsymbol{X}) = Yp(\boldsymbol{X}) + (1 - Y)(1 - p(\boldsymbol{X}))$$

where $p(\boldsymbol{X})$ naturally represents

$$p(\boldsymbol{X}) = P(Y = 1|\boldsymbol{X}) .$$

So the sample distribution will be

$$P(Y_i, \boldsymbol{X}_i) = p(Y_i|\boldsymbol{X}_i)p_0(\boldsymbol{X}_i) .$$

If the sample $\{\boldsymbol{X}_i\}$ is drawn with a random sampling from the total image consisting of N_{tot} pixels, we have

$$p_0(\boldsymbol{X}_i) = \frac{1}{N_{\text{tot}}} .$$

Of course, it is possible to adopt finer sampling strategies, but we limit ourselves to this in the example.

Having to consider the logarithm of $p(Y_i, \boldsymbol{X}_i)$, $-\log N_{\text{tot}}$ becomes an additive constant that therefore does not influence the optimization, so we neglect this factor. As you can see, the problem of estimating $p(Y, \boldsymbol{X})$ is reduced to that of estimating the conditional $p(\boldsymbol{X})$. This will be approximated by the family $q(\boldsymbol{X}|\boldsymbol{\vartheta})$, with the caveat that the value of q is already between 0 and 1. We can then draw a very simple neural network as in Fig. 6.5, assuming in this case that the activation function between h and Y this time is a sigmoid that returns a number between 0 and 1.

We note that unlike the previous figures, we have also represented the units X_0 and h_0 here, which always assume the value 1 for each sample value.

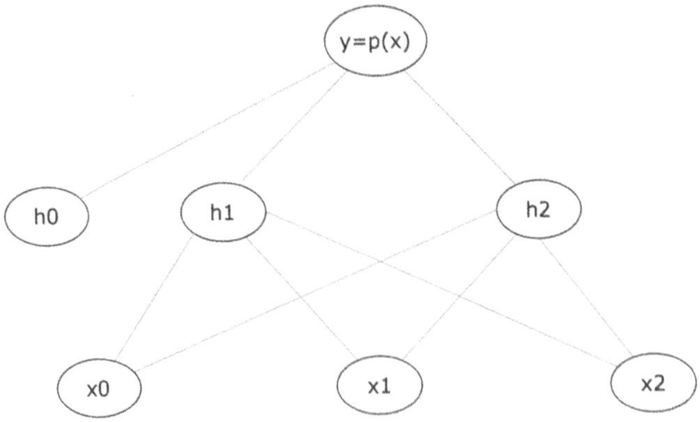

Fig. 6.5 The NN for binary classification

The analytical expression of $q(\boldsymbol{h}|\boldsymbol{\vartheta})$ then becomes

$$q(\boldsymbol{h}|\boldsymbol{\vartheta}) = \sigma(\boldsymbol{\gamma}^\top \boldsymbol{h} + \gamma_0)$$
$$h_1(\boldsymbol{X}|\boldsymbol{\vartheta}) = \sigma(\boldsymbol{\alpha}_1^\top \boldsymbol{X} + \alpha_{01})$$
$$h_2(\boldsymbol{X}|\boldsymbol{\vartheta}) = \sigma(\boldsymbol{\alpha}_2^\top \boldsymbol{X} + \alpha_{02})$$
$$\boldsymbol{\gamma} = \begin{vmatrix} \gamma_1 \\ \gamma_2 \end{vmatrix}$$
$$\boldsymbol{h} = \begin{vmatrix} h_1 \\ h_2 \end{vmatrix}$$
$$\boldsymbol{\alpha}_1 = \begin{vmatrix} \alpha_{11} \\ \alpha_{12} \end{vmatrix}, \ \boldsymbol{\alpha}_2 = \begin{vmatrix} \alpha_{21} \\ \alpha_{22} \end{vmatrix}$$

with

$$\boldsymbol{\vartheta}^\top = [\gamma_0, \boldsymbol{\gamma}^\top, \alpha_{01}, \boldsymbol{\alpha}_1^\top, \alpha_{02}, \boldsymbol{\alpha}_2^\top] .$$

So we can write

$$q_i = q(Y_i, \boldsymbol{X}_i|\boldsymbol{\vartheta}) = Y_i \sigma(\gamma_0 + \boldsymbol{\gamma}^\top \boldsymbol{h}_i) + (1 - Y_i)[1 - \sigma(\gamma_0 + \boldsymbol{\gamma}^\top \boldsymbol{h}_i)]$$

where

$$\boldsymbol{h}_i = \begin{vmatrix} \sigma(\boldsymbol{\alpha}_1^\top \boldsymbol{X}_i + \alpha_{01}) \\ \sigma(\boldsymbol{\alpha}_2^\top \boldsymbol{X}_i + \alpha_{02}) \end{vmatrix} .$$

We also observe that, due to the binary nature of Y_i (0,1), we can write

$$\log q(Y_i, \boldsymbol{X}_i|\boldsymbol{\vartheta}) = Y_i \log \sigma(\gamma_0 + \boldsymbol{\gamma}^\top \boldsymbol{h}_i) + (1 - Y_i) \log[1 - \sigma(\gamma_0 + \boldsymbol{\gamma}^\top \boldsymbol{h}_i]$$

with $\boldsymbol{h}_i$ as above.

So the function to optimize will be

$$K_\lambda(\boldsymbol{\vartheta}) = -\frac{1}{N} \sum \{Y_i \log \sigma(\gamma_0 + \boldsymbol{\gamma}^\top \boldsymbol{h}_i) + (1 - Y_i) \log[1 - \sigma(\gamma_0 + \boldsymbol{\gamma}^\top \boldsymbol{h}_i)]\} +$$
$$+\lambda\{\gamma_0^2 + |\boldsymbol{\gamma}|^2 + \alpha_{01}^2 + |\boldsymbol{\alpha}_1|^2 + \alpha_{02}^2 + |\boldsymbol{\alpha}_2|^2\} .$$

The example is easily generalized to the case where the input variables $\{X_k\}$ are actually vectors, as happens for example in remote sensing, and the number of classes in the classification is any.

We now see how to use a NN to optimize $J_\lambda(\boldsymbol{\vartheta})$ or $K_\lambda(\boldsymbol{\vartheta})$. For this purpose we will use the NN in two modes: feed forward and backward propagation. These, from a graphical point of view, correspond to orienting the connections of the NN graph from a lower level to the next or vice versa.

The feed forward mode essentially corresponds to the calculation of Y from the inputs $\{X_k\}$; for a vanilla neural network it is to calculate $\boldsymbol{h}$ from $\{X_k\}$ and subsequently Y from $\boldsymbol{h}$ with the transformations fixed in the definition of the NN. In regression this serves to calculate $g(\boldsymbol{X}|\boldsymbol{\vartheta})$, for a fixed $\boldsymbol{\vartheta}$, and therefore has two possible uses: in the context of optimization to calculate $g(\boldsymbol{X}_i|\boldsymbol{\vartheta})$ and then $v_i = Y_i - g(\boldsymbol{X}_i|\boldsymbol{\vartheta})$ which serve to calculate the $\nabla_{\boldsymbol{\vartheta}} g(\boldsymbol{X}_i|\boldsymbol{\vartheta})$ necessary to take a step in the succession of the steepest descent; otherwise $g(\boldsymbol{X}|\boldsymbol{\vartheta})$ can become the predicted value of the field in question at a point $\boldsymbol{X}$ where we do not have sample data. Of course in this case the $\boldsymbol{\vartheta}$ to use in $g(\boldsymbol{X}|\boldsymbol{\vartheta})$ must be the one that results from the optimization process. The same applies to the calculation of the distribution $Q(Y, \boldsymbol{X}|\boldsymbol{\vartheta})$. In particular the prediction of $Q(Y, \boldsymbol{X}|\boldsymbol{\vartheta})$ when Y is an integer valued variable, as in the classification problem, serves to predict the most probable value of the class on the basis of the principle of maximum likelihood

$$k(\boldsymbol{X}) = \arg\max_{\ell} Q(Y = \ell, \boldsymbol{X}|\boldsymbol{\vartheta}) \,. \tag{6.41}$$

The backward propagation mode of using the NN corresponds to the calculation of $\nabla_{\vartheta} J_\lambda(\boldsymbol{\vartheta})$ or of $\nabla_{\vartheta} K_\lambda(\boldsymbol{\vartheta})$, depending on the case.

The calculation is essentially elementary and uses the mechanism of the product of derivatives of function of function. Rather than giving a general rule, we present here the completion of Examples 6.3 and 6.4, which show how to arrive at the calculation of the gradient for both $J_\lambda(\boldsymbol{\vartheta})$ and $K_\lambda(\boldsymbol{\vartheta})$.

Example 6.5 This example aims to complete Example 6.3 by showing its backward propagation algorithm. We recall that in this example the input units contain vectors $\boldsymbol{X}_{ki}$, $k = 1, \ldots n$; $i = 1 \ldots N$, rather than scalars; to these is added the input $X_{0i} \equiv 1$ which is scalar.

We observe that a feed forward cycle provides us, for a given $\boldsymbol{\vartheta}$, the values of Table 6.1.

Table 6.1 Variables calculated by the feed forward

Input level:	$X_{0i} \equiv 1\ ,\ \boldsymbol{X}_{1i}\ ,\ \boldsymbol{X}_{2i}\ ,\ \boldsymbol{X}_{3i}$
Hidden level:	$h_{0i} \equiv 1,$
	$h_{1i} = \sigma(\sum_{k=1}^{3} \boldsymbol{\alpha}_{k1}^T \boldsymbol{X}_{ki} + \alpha_{01} X_{0i})$
	$h_{2i} = \sigma(\sum_{k=1}^{3} \boldsymbol{\alpha}_{k2}^T \boldsymbol{X}_{ki} + \alpha_{02} X_{0i})$
	$h_{3i} = \sigma(\sum_{k=1}^{3} \boldsymbol{\alpha}_{k3}^T \boldsymbol{X}_{ki} + \alpha_{03} X_{0i})$
Output level:	$\widehat{Y}_i = \sum_{\ell=1}^{3} \gamma_\ell h_{\ell i} + \gamma_0 h_{0i}$

Table 6.2 Values derived from Table 6.1

$\sigma_{\ell i} = \sigma(\sum_{k=1}^{3} \boldsymbol{\alpha}_{k\ell}^{\top} \boldsymbol{X}_{ki} + \alpha_{0\ell}) = h_{\ell i} \quad (\ell = 1, 2, 3)$
$\sigma'_{\ell i} = h_{\ell i}(1 - h_{\ell i})$
$v_i = Y_i - \widehat{Y}_i = Y_i - (\sum_{\ell=1}^{3} \gamma_\ell h_{\ell i} + \gamma_0)$

In Table 6.2 we observe that σ is always a function of a single scalar variable and also the identity holds

$$\sigma'(t) = \sigma(t)(1 - \sigma(t)) .$$

Therefore, the derived quantities of Table 6.2 can be calculated too from the values of Table 6.1.

From these values it is easy to find the required gradients, namely in order (backward)

$$\frac{\partial J_\lambda(\boldsymbol{\vartheta})}{\partial \gamma_0}, \frac{\partial J_\lambda(\boldsymbol{\vartheta})}{\partial \gamma_\ell} \frac{\partial J_\lambda(\boldsymbol{\vartheta})}{\partial \alpha_{0\ell}}, \nabla_{\boldsymbol{\alpha}_{k\ell}} J_\lambda(\boldsymbol{\vartheta}) . \tag{6.42}$$

In fact, remembering that

$$J_\lambda(\boldsymbol{\vartheta}) = \frac{1}{N} \sum_{i=1}^{N} v_i^2 + \lambda |\boldsymbol{\vartheta}|^2$$

we obtain

$$\frac{\partial J_\lambda(\boldsymbol{\vartheta})}{\partial \gamma_0} = -\frac{2}{N} \sum_{i=1}^{N} v_i + 2\lambda\gamma_0$$

$$\frac{\partial J_\lambda(\boldsymbol{\vartheta})}{\partial \gamma_\ell} = -\frac{2}{N} \sum_{i=1}^{N} v_i h_{\ell i} + 2\lambda\gamma_\ell$$

$$\frac{\partial J_\lambda(\boldsymbol{\vartheta})}{\partial \alpha_{0\ell}} = -\frac{2}{N} \sum_{i=1}^{N} v_i \frac{\partial v_i}{\partial \alpha_{0\ell}} + 2\alpha_{0\ell} = -\frac{2}{N} \sum_{i=1}^{N} v_i \gamma_\ell \sigma'_{\ell i} + 2\alpha_{0\ell}$$

$$\nabla_{\boldsymbol{\alpha}_{k\ell}} J_\lambda(\boldsymbol{\vartheta}) = -\frac{2}{N} \sum_{i=1}^{N} v_i \nabla_{\boldsymbol{\alpha}_{k\ell}} v_i + 2\boldsymbol{\alpha}_k = -\frac{2}{N} \sum_{i=1}^{N} v_i \gamma_\ell \sigma'_{\ell i} \boldsymbol{X}_{ki} + 2\boldsymbol{\alpha}_k .$$

Therefore, given a value of $\boldsymbol{\vartheta}$ (and of λ), a feed forward-backward propagation cycle is able to calculate the $\nabla_{\boldsymbol{\vartheta}} J_\lambda(\boldsymbol{\vartheta})$, which is what we need to take a step of optimization according to the steepest descent.

The further concatenation with the variation of λ is described in Appendix D.

Example 6.6 In this example we see the backward propagation cycle related to Example 6.4, classification into two classes. Therefore, we use the same notation as in Example 6.4. Our aim is to calculate the $\nabla_{\boldsymbol{\vartheta}} K_\lambda(\boldsymbol{\vartheta})$, with

$$K_\lambda(\boldsymbol{\vartheta}) = -\frac{1}{N}\sum_{i=1}^{N} \log q_i + \lambda|\boldsymbol{\vartheta}|^2 ,$$

or

$$\nabla_{\boldsymbol{\vartheta}} K_\lambda(\boldsymbol{\vartheta}) = -\frac{1}{N}\sum_{i=1}^{N} \nabla_{\boldsymbol{\vartheta}} \log q_i + 2\lambda\boldsymbol{\vartheta} .$$

Therefore, we need to find $\nabla_{\boldsymbol{\vartheta}} \log q_i$, where the parameter vector is

$$\boldsymbol{\vartheta}^\top = [\gamma_0, \boldsymbol{\gamma}^\top, \alpha_{01}, \boldsymbol{\alpha}_1^\top, \alpha_{02}, \boldsymbol{\alpha}_2^\top] .$$

and also

$$\log q_i = Y_i \log\sigma(\gamma_0 + \boldsymbol{\gamma}^\top \boldsymbol{h}_i) + (1 - Y_i)\log[1 - \sigma(\gamma_0 + \boldsymbol{\gamma}^\top \boldsymbol{h}_i)] .$$

Remember that

$$\boldsymbol{X}_i = \begin{vmatrix} X_{1i} \\ X_{2i} \end{vmatrix} , \quad \boldsymbol{h}_i = \begin{vmatrix} h_{1i} \\ h_{2i} \end{vmatrix} \equiv \begin{vmatrix} \sigma(\boldsymbol{\alpha}_1^\top \boldsymbol{X}_i + \alpha_{01}) \\ \sigma(\boldsymbol{\alpha}_2^\top \boldsymbol{X}_i + \alpha_{02}) \end{vmatrix} .$$

For convenience, let's set

$$m_{\ell i} = \boldsymbol{\alpha}_\ell^\top \boldsymbol{X}_i + \alpha_{0i} ,$$
$$k_i = \gamma_0 + \boldsymbol{\gamma}^\top \boldsymbol{h}_i .$$

so that

$$\log q_i = Y_i \log\sigma(k_i) + (1 - Y_i)\log[1 - \sigma(k_i)]$$

and

$$\boldsymbol{h}_i = \begin{vmatrix} \sigma(m_{1i}) \\ \sigma(m_{2i}) \end{vmatrix} .$$

Furthermore, we observe that, from $\sigma'(t) = \sigma(1 - \sigma)$ we get

$$[\log\sigma(t)]' = 1 - \sigma(t) , \quad [\log(1 - \sigma(t))]' = -\sigma(t) .$$

As in the previous exercise, a feed forward cycle provides us with all the numerical values of the input quantities, of $m_{\ell i}, k_i, \boldsymbol{h}_i$ and q_i, when $\boldsymbol{\vartheta}$ has been set.

So we can proceed to calculate

$$\frac{\partial \log q_i}{\partial \gamma_0} = \frac{\partial \log q_i}{\partial k_i}\frac{\partial k_i}{\partial \gamma_0} = Y_i - \sigma(k_i)\,,$$

$$\nabla_{\gamma} \log q_i = \frac{\partial \log q_i}{\partial k_i}\nabla_{\gamma} k_i = [Y_i - \sigma(k_i)]\boldsymbol{h}_i\,;$$

and again

$$\begin{aligned}
\frac{\partial \log q_i}{\partial \alpha_{0\ell}} &= \frac{\partial \log q_i}{\partial k_i}\frac{\partial k_i}{\partial \alpha_{0\ell}} = \\
&= [Y_i - \sigma(k_i)]\sum_{i=1}^{2} \gamma_j \frac{\partial h_{ji}}{\partial \alpha_{0\ell}} \\
&= [Y_i - \sigma(k_i)]\gamma_\ell \sigma'(m_{\ell i})\frac{\partial (m_{\ell i})}{\partial \alpha_{0\ell}} = \\
&= [Y_i - \sigma(k_i)]\gamma_\ell \sigma(m_{\ell i})[1 - \sigma(m_{\ell i})]\,,
\end{aligned}$$

$$\begin{aligned}
\nabla_{\alpha_\ell} \log q_i &= \frac{\partial \log q_i}{\partial k_i}\nabla_{\alpha_\ell} k_i = [Y_i - \sigma(k_i)]\sum_{j=1}^{2} \gamma_j \nabla_{\alpha_\ell} h_{ji} = \\
&= [Y_i - \sigma(k_i)]\gamma_\ell \sigma'(m_\ell)\nabla_{\alpha_\ell} m_{\ell i} = \\
&= [Y_i - \sigma(k_i)]\gamma_\ell \sigma(m_{\ell i})[1 - \sigma(m_{\ell i})]\boldsymbol{X}_i\,.
\end{aligned}$$

Therefore, a feed forward-backward propagation cycle allows the calculation of $\nabla_{\vartheta} K_\lambda(\boldsymbol{\vartheta})$, which is what we need to take a step in minimizing $K_\lambda(\boldsymbol{\vartheta})$.

With a clear anthropomorphic allusion, the phase of model creation and estimation of the optimal values of $\boldsymbol{\vartheta}$ and λ, as described in this section and in the previous ones, is referred to in the literature as training the NN, or learning by the NN. Once the model is validated, assessing its quality based on the test sample, the NN is ready to be used in prediction mode, that is to calculate:

- for regression $\widehat{Y}(\boldsymbol{X}) = \widehat{Y}(\boldsymbol{X}|\widehat{\boldsymbol{\vartheta}}, \widehat{\lambda})$
- for classification $\widehat{k}(\boldsymbol{X}) = \arg\max_h \widehat{p}_h(\boldsymbol{X}|\widehat{\boldsymbol{\vartheta}}, \widehat{\lambda})$

for any value of $\boldsymbol{X}$ not included in the initial sample.

We conclude this section by noting that beyond the formulas and the simplicity of the examples presented, deep machine learning presents itself as a set of algorithms that implement machine learning based on the optimization of empirical models that contain a large number of parameters, controlled by appropriate regularization. Of course, this approach exposes itself to the risk of creating models that contradict the known physics of the phenomena to be described/predicted. Therefore, the results, beyond the mystique of artificial intelligence, must be validated by an expert in the application area.

Even more, experts should be consulted before creating the statistical model in such a way as to include in it "simple", well known features of the experiment under analysis. Quoting Cox and Hinkley [7] "It is therefore worth stressing that very simple theoretical analyses can be valuable as a basis of statistical analysis".

Chapter 7
Some Conclusions

The main concepts referring to the role of probabilistic theory and statistics in improving our knowledge of physical world, with a focus on Earth Sciences, are summarized.

7.1 Learning or Understanding?

The first conclusion we want to draw from the material presented in the text is that the "probabilistic vision of the physical world" is an indispensable tool for understanding and modeling the world in terms of physical laws. This view of the world is indeed the space in which the deterministic, mechanistic view, and the purely empirical view, represented by the black box theory, coexist, united by a process that selects the stable behaviors of the phenomena of physical reality; they are then represented in the laws of nature distinct from those uncertain, which therefore present themselves with a degree of indeterminacy describable with a probabilistic model, and finally from those that we represent with a form of complete ignorance of a part that enters into the observation equations of the observed physical phenomenon, dominated by the absence of cause-effect.

Naturally, this behavior also has its probabilistic modeling, the archetype of which is the concept of white noise.

In this sense, we believe that a progress of scientific knowledge is realized within the probabilistic vision of the world, contrary to what Lindley [29] asserts, for whom the job of the statistician is to represent the uncertainty of the "client", for example a scientist, in terms of probability.

A second conclusion is that machine learning algorithms do indeed allow the calculation of optimization, but the results must always be checked to avoid contradicting well-known physical laws, which, when possible, must be used a priori in building the statistical model. But this vision, the mathematics and

F. Sansò, A. Albertella, *The Probabilistic Vision of the Physical World*, Lecture Notes in Geosystems Mathematics and Computing,
https://doi.org/10.1007/978-3-031-88268-5_7

algorithms that support it, is itself sufficient, even through an enormous increase in data and a huge increase in computing capacity, to autonomously create a coherent model of physical reality, even just within the field of Earth sciences?

In other words, can artificial intelligence not only be a useful tool but an autonomous entity capable of substituting the work of the scientist in understanding and constructing the physics of the Earth, or more generally the physics of the Universe?

We believe not. Returning to the authoritative voice of Karl Popper [39], we find that the author develops an idea of the evolution of knowledge, believing that its progress is based on the sequential elimination of erroneous theories and the approximation of truth. Popper represents this idea in a tetradic formula.

$$P_1 - TT - EE - P_2\,,$$

that is, initial problem, tentative theory, error elimination through critical analysis, emergence of a new problem, adding that a chain of cycles of this type tends towards truth.

We agree with the first part of the statement, on advancement made by progressively eliminating erroneous theories, or perhaps not adequate in a context

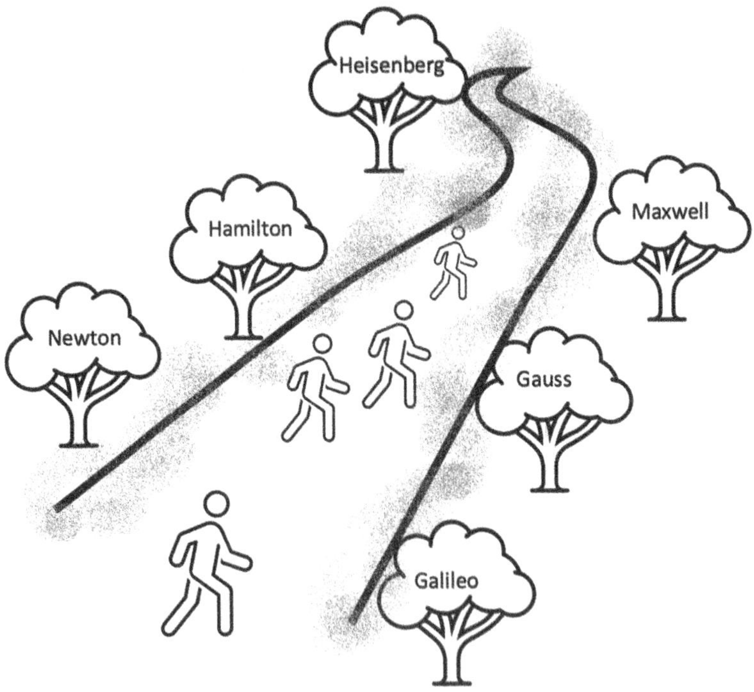

Fig. 7.1 The road of knowledge of the physical world

of expanding data, but we disagree with the second part, which presupposes the objective and demonstrated existence of an immutable truth towards which we tend.

In reality, we believe, for what we can understand, that there is a domain of regularity of the results of experiments, that is, of the interaction between man and the physical world, and that its connection with objectivity can only be mediated by a probabilistic model, precisely the "probabilistic vision of the physical world". The knowledge thus acquired appears to us as a road that we travel in one direction and the other; the ends of the road are hidden by the fog of indeterminacy, but this does not make the piece of road we see less real: its reality and its rules cannot be ignored, but must be conquered meter by meter (Fig. 7.1).

In support of our point of view, we again cite M.Planck, in his preface to the text "The Knowledge of the Physical World" of 1943 [38]. The author writes: "So it happens also for what concerns the real world, which in the end is not the starting point, but the final goal of physical research, a goal that can never be completely reached, but must always be kept in mind if one wants to proceed forward".

Appendix A
Some Recalls of Probability Theory

Given a probability space $(\Omega, \mathcal{A}, P)$ (see Sect. 1.1), by definition the conditional probability of event A given an event B, when $P(B) > 0$, is set as

$$P(A|B) = \frac{P(A \cap B)}{P(B)} . \tag{A.1}$$

When B has zero probability, assuming that Ω possesses a metric, for example is a subset of $\mathbb{R}^n$, defining the expansion of B as

$$B_\varepsilon \equiv \{\omega \ , \ d(\omega, B) \leq \varepsilon\} \tag{A.2}$$

one can set, when the limit exists,

$$P(A|B) = \lim_{\varepsilon \to 0} \frac{P(A \cap B_\varepsilon)}{P(B_\varepsilon)} . \tag{A.3}$$

Theorem A.1 (Bayes' Theorem) *Let $\{A_i\}$ be a disjoint decomposition of Ω, that is*

$$A_i \in \mathcal{A}, \ \bigcup_{i=1}^{N} A_i = \Omega \ , \ A_i \cap A_k = \emptyset \quad i \neq k \ , \tag{A.4}$$

then it holds

$$P(A_i|B) = \frac{P(A_i \cap B)}{\sum_{k=1}^{N} P(A_k \cap B)} \tag{A.5}$$

F. Sansò, A. Albertella, *The Probabilistic Vision of the Physical World*, Lecture Notes in Geosystems Mathematics and Computing,
https://doi.org/10.1007/978-3-031-88268-5

Defining

$$x, a \in \mathbb{R}^n \ , \quad \{x \leq a\} \Leftrightarrow x_i \leq a_i \quad i = 1 \dots N \ , \tag{A.6}$$

we recall that a random variable (or r.v.) $X \in \mathbb{R}^n$ is a measurable function of ω, that is the sets

$$\{\omega \ ; \ X(\omega) \leq a\} \tag{A.7}$$

are in $\mathcal{A}$, $\forall a \in \mathbb{R}^n$.

If $\Omega \subset \mathbb{R}^n$ one can directly set

$$x = \omega \tag{A.8}$$

being careful however that in this way we extend the probability distribution from Ω to all $\mathbb{R}^n$, beyond the conceptual difference for which $x \in \Omega^c$ is an impossible event, since in this way it will be

$$P(\Omega^c) = 0 \ ; \tag{A.9}$$

so all probabilistic calculations are not altered and the (A.8) becomes acceptable.

We define two events as stochastically independent if

$$P(A \cap B) = P(A)P(B) \ ; \tag{A.10}$$

it is clear that for two independent events

$$P(A|B) = P(A) \tag{A.11}$$

and vice versa, therefore also

$$P(B|A) = P(B) \ . \tag{A.12}$$

Similarly, for a double r.v. (X, Y) we say that X and Y are stochastically independent if

$$\forall A_1, A_2 \subset \mathbb{R} \quad P((X, Y) \in A_1 \times A_2) = P(X \in A_1)P(Y \in A_2) \ , \tag{A.13}$$

where $A_1 \times A_2$ is the Cartesian product of A_1 with A_2. Equation (A.13) implies that for a discrete r.v. with argumental values (x_i, y_k) we have

$$P(X = x_i \ , \ Y = y_k) = P(X = x_i)P(Y = y_k) \ , \tag{A.14}$$

while for a regular continuous r.v., characterized by a probability density $f(x, y)$ (recall the definition (1.17)), it must hold

$$f(x, y) = f_X(x) f_Y(y) , \tag{A.15}$$

where $f_X(x)$, $f_Y(y)$ are the marginals of $f(x, y)$, that is

$$f_X(x) = \int f(x, y) dy , \quad f_Y(y) = \int f(x, y) dx . \tag{A.16}$$

These concepts can be easily generalized to a r.v. $\boldsymbol{X}$ in $\mathbb{R}^n$.

The concept of conditional probability can also be brought back, for example for a double variable (X, Y), to that of the distribution of one variable, for example Y, conditioned on the value of the other, for example X.

For a discrete variable we will have

$$P(Y = y_k | X = x_i) = \frac{P(X = x_i \, , Y = y_k)}{P(X = x_i)} , \tag{A.17}$$

while for a regular continuous variable the relationship between probability densities holds (see (1.17))

$$f(Y|X) = \frac{f(x, y)}{f_X(x)} . \tag{A.18}$$

Note that for every x, Eq. (A.18) gives a probability density for a variable Y, as is also evident from the fact that integrating both members in dy we find

$$\int f(y|x) dy \equiv 1 \;\; \forall x, \; f_X(x) > 0 . \tag{A.19}$$

Focusing on continuous r.v. in $\mathbb{R}^n$, we recall the definitions of mean and covariance,

$$\mu_X = E\{\boldsymbol{X}\} = \int \boldsymbol{x} f(\boldsymbol{x}) d_n x , \tag{A.20}$$

$$C = E\{(\boldsymbol{X} - \mu_X)(\boldsymbol{X} - \mu_X)^{\top}\} . \tag{A.21}$$

In particular, the covariance is a symmetric and positive definite matrix; if the variable $\boldsymbol{X}$ is regular, C is strictly positive definite, and therefore invertible.

On the diagonal of C we have the variances of the components X_i

$$C_{ii} = \sigma^2(X_i) = E\{(X_i - \mu_{X_i})^2\} . \tag{A.22}$$

When the components are independent

$$\begin{aligned} i \neq k \qquad C_{ik} &= E\{(X_i - \mu_{X_i})(X_k - \mu_{X_k})\} = \\ &= E\{X_i - \mu_{X_i}\} \cdot E\{X_k - \mu_{X_k}\} = 0 . \end{aligned} \tag{A.23}$$

Conversely, if Eq. (A.23) holds, X_i is said to be *linearly independent* or *uncorrelated* with X_k; this does not generally imply stochastic independence. It is clear that μ_X is one of the so-called position indices of the distribution $f(\boldsymbol{x})$, while C is a dispersion index, as can be seen from (A.22).

If the variable $\boldsymbol{X} \in \mathbb{R}^n$ is linearly transformed into a variable $\boldsymbol{Y} \in \mathbb{R}^m$, that is

$$\boldsymbol{Y} = A\boldsymbol{X} + \boldsymbol{a} , \tag{A.24}$$

mean and covariance transform according to the laws

$$\mu_Y = A\mu_X + \boldsymbol{a} \tag{A.25}$$

$$C_Y = AC_X A^\top . \tag{A.26}$$

Equation (A.26) is called *covariance propagation*. When the components of $\boldsymbol{X}$ are independent and $Y \in \mathbb{R}^1$, that is

$$Y = \sum_{i=1}^{n} a_i X_i , \tag{A.27}$$

Eq. (A.26) gives

$$\sigma_Y^2 = \sum_{i=1}^{n} a_i^2 \sigma_{X_i}^2 \tag{A.28}$$

which is called *variance propagation*.

Recalling Eq. (A.18) it is possible to define the mean of the conditional distribution $f(\boldsymbol{y}|\boldsymbol{x})$ for each $\boldsymbol{x}$; thus we find a function of $\boldsymbol{x}$ also called *regression curve* of $\boldsymbol{y}$ on $\boldsymbol{x}$;

$$\widehat{\boldsymbol{y}}(\boldsymbol{x}) = E\{\boldsymbol{Y}|\boldsymbol{X} = \boldsymbol{x}\} = \int \boldsymbol{y} f(\boldsymbol{y}|\boldsymbol{x}) d\boldsymbol{y} . \tag{A.29}$$

Clearly $E\{\boldsymbol{Y}|\boldsymbol{X} = \boldsymbol{x}\}$ is a function of $\boldsymbol{x}$; if now, instead of fixing $\boldsymbol{X}$ to a value, we let it vary as a random variable, we obtain another random variable function of $\boldsymbol{X}$

$$E\{\boldsymbol{Y}|\boldsymbol{X}\} = \int \boldsymbol{y} f(\boldsymbol{y}|\boldsymbol{X}) d\boldsymbol{y} ; \tag{A.30}$$

Eq. (A.30) is called *regression variable* of $\boldsymbol{Y}$ on $\boldsymbol{X}$ and sometimes indicated as $\boldsymbol{Y}|\boldsymbol{X}$ or $\widehat{Y}(\boldsymbol{X})$.

An important development of the concept of conditional probability is given by the fundamental Bayes' theorem, which we prove here for continuous and regular variables.

Theorem A.2 (Bayes' Theorem) *The following holds*

$$f(y|x) = \frac{f(x|y)f_Y(y)}{\int f(x|\eta)f_Y(\eta)d\eta} . \tag{A.31}$$

Indeed, from the definition (A.18) and the identity

$$f(x, y) = f(y|x)f_X(x) = f(x|y)f_Y(y) \tag{A.32}$$

we obtain

$$f(y|x) = \frac{f(x, y)}{f_X(x)} = \frac{f(x|y)f_Y(y)}{\int f(x, \eta)d\eta} = \frac{f(x|y)f_Y(y)}{\int f(x|\eta)f_Y(\eta)d\eta} . \tag{A.33}$$

Recalling (A.30), we find a corollary for the regression variable of $\boldsymbol{Y}$ on $\boldsymbol{X}$.

Corollary A.3 *For the regression variable of* $\boldsymbol{Y}$ *on* $\boldsymbol{X}$*, it results*

$$E\{\boldsymbol{Y}|\boldsymbol{X}\} = \widehat{\boldsymbol{Y}}(\boldsymbol{X}) = \frac{\int \boldsymbol{y} f(\boldsymbol{X}|\boldsymbol{Y})f_Y(\boldsymbol{y})d\boldsymbol{y}}{f_X(\boldsymbol{x})} = \frac{\int \boldsymbol{y} f(\boldsymbol{X}|\boldsymbol{Y})f_Y(\boldsymbol{y})d\boldsymbol{y}}{\int f(\boldsymbol{X}|\boldsymbol{Y})f_Y(\boldsymbol{y})d\boldsymbol{y}} . \tag{A.34}$$

The interpretation of Bayes' theorem is discussed in Chap. 3.

It is useful to briefly demonstrate here the fundamental theorem of stochastic approximation.

Theorem A.4 *Given any* $\boldsymbol{h}(\boldsymbol{X}) \in \mathbb{R}^m$*, we define the mean square error of the prediction of* $\boldsymbol{Y}$ *with* $\boldsymbol{h}(\boldsymbol{x})$ *as*

$$\mathcal{E}^2(\boldsymbol{h}) = E\{|\boldsymbol{Y} - \boldsymbol{h}(\boldsymbol{X})|^2\} = \int |\boldsymbol{y} - \boldsymbol{h}(\boldsymbol{x})|^2 f(\boldsymbol{x}, \boldsymbol{y})d_n x d_m y ; \tag{A.35}$$

when the integral is finite. It results

$$\mathcal{E}^2(\boldsymbol{h}) \geq \mathcal{E}^2(\widehat{\boldsymbol{Y}}(\boldsymbol{X})) \tag{A.36}$$

with $\widehat{\boldsymbol{Y}}(\boldsymbol{X})$ *given by (A.34), that is* $\mathcal{E}^2(\boldsymbol{h})$ *is minimum in* $\widehat{\boldsymbol{Y}}(\boldsymbol{X})$.

In fact, it is enough to observe that whatever $\boldsymbol{g}(\boldsymbol{x})$ is, provided it has finite variance,

$$E\{[\boldsymbol{Y} - \widehat{\boldsymbol{Y}}(\boldsymbol{X})]^\top \boldsymbol{g}(\boldsymbol{X})\} = \int [\boldsymbol{Y} - \widehat{\boldsymbol{Y}}(\boldsymbol{x})]^\top \boldsymbol{g}(\boldsymbol{x}) f(\boldsymbol{x}, \boldsymbol{y})d_n x d_m y =$$

$$= \int d_n x\, \boldsymbol{g}^\top(\boldsymbol{x}) \int d_m\, y[\boldsymbol{Y} - \widehat{\boldsymbol{Y}}(\boldsymbol{x})] f(\boldsymbol{x}, \boldsymbol{y}) = 0 \tag{A.37}$$

because, for (A.34), (A.16),

$$\int d_m y[\boldsymbol{Y} - \widehat{\boldsymbol{Y}}(\boldsymbol{x})] f(\boldsymbol{x}, \boldsymbol{y}) = \widehat{\boldsymbol{Y}}(\boldsymbol{x}) f_X(\boldsymbol{x}) - \widehat{\boldsymbol{Y}}(\boldsymbol{x}) f_X(\boldsymbol{x}) = 0 \,. \tag{A.38}$$

Therefore, we have

$$\begin{aligned} \mathcal{E}^2(\boldsymbol{h}) &= E\{|\boldsymbol{Y} - \widehat{\boldsymbol{Y}}(\boldsymbol{X})|^2\} + 2E\{[\boldsymbol{Y} - \widehat{\boldsymbol{Y}}(\boldsymbol{X})]^\top [\widehat{\boldsymbol{Y}}(\boldsymbol{X}) - \boldsymbol{h}(\boldsymbol{X})]\} + \\ &+ E\{|\widehat{\boldsymbol{Y}}(\boldsymbol{X}) - \boldsymbol{h}(\boldsymbol{X})|^2\} = \mathcal{E}^2((\widehat{\boldsymbol{Y}})) + E\{|\widehat{\boldsymbol{Y}}(\boldsymbol{X}) - \boldsymbol{h}(\boldsymbol{X})|^2\} \,, \end{aligned} \tag{A.39}$$

since the intermediate term is null for (A.37); (A.36) is proven.

We close this appendix of recalls with a brief discussion of families of normal or Gaussian variables.

A random variable X in $\mathbb{R}^1$ is said to be normal if its probability density is given by

$$f(x) = \frac{1}{\sqrt{2\pi}\sigma} e^{-\frac{1}{2}\frac{(x-\mu)^2}{\sigma^2}} \,. \tag{A.40}$$

It is easy to see that in this case

$$E\{X\} = \mu \tag{A.41}$$

and

$$\sigma^2(X) = E\{(X - \mu)^2\} = \sigma^2 \,. \tag{A.42}$$

Equation (A.40) is often represented with the formula

$$X \sim \mathcal{N}(\mu, \sigma^2) \,. \tag{A.43}$$

In the case that $\mu = 0$ and $\sigma = 1$, we speak of standardized normal, $\mathcal{Z}$.

The importance of this distribution lies in the fact that it represents the limit distribution of many sequences of random variables. In this regard, there are several theorems, but here we limit ourselves to the formulation of Lyapunov [32] adapted to our case; for a deeper study, see [10], Chap. 2, §3.

Theorem A.5 (Central Limit Theorem (CLT)) *Let $\{\xi_n\}$ be a sequence of stochastically independent variables such that*

$$E\{\xi_n\} = 0 \,, \; \sigma^2(\xi_n) = E\{\xi_n^2\} = \sigma_n^2 \,, \; E\{|\xi_n|^3\} \leq c \,; \tag{A.44}$$

let us put

$$\begin{aligned} X_N &= \sum_{n=1}^{N} \xi_n \\ s_N^2 &= \sum_{n=1}^{N} \sigma_n^2 = \sigma^2(X_N) \\ Z_N &= \frac{X_N}{s_N}\,; \end{aligned} \tag{A.45}$$

then it follows that the distribution $f_{Z_N}(z)$ *tends to that of the standardized normal* $f_{\mathcal{Z}}(z)$, *in the sense that* $\forall[a, b]$

$$\int_a^b f_{Z_N}(t)dt \to \int_a^b f_Z(t)dt\,. \tag{A.46}$$

It follows that for large N it will be

$$f_{Z_N}(z) \cong f_{\mathcal{Z}}(z)\,, \tag{A.47}$$

or

$$f_{X_N}(x) \cong \mathcal{N}(0, s_N^2)\,. \tag{A.48}$$

It is useful to note here that when the variances σ_n^2 are bounded,

$$\sigma_n^2 \leq \sigma_0^2\,, \tag{A.49}$$

it will be

$$s_N^2 \leq N\sigma_0^2\,, \tag{A.50}$$

so that setting

$$\overline{X}_N = \frac{1}{N}\sum_{n=1}^{N} \xi_n = \frac{1}{N} X_N \tag{A.51}$$

we have

$$\sigma^2(\overline{X}_N) \leq \frac{N\sigma_0^2}{N^2} = \frac{\sigma_0^2}{N} \xrightarrow[N\to\infty]{} 0\,; \tag{A.52}$$

that is, the averages $\overline{X}_N$ tend to zero in quadratic mean and therefore in probability. This is the so-called weak law of large numbers. We also note that if $E\{\xi_n\} = \mu_n$

and $\frac{1}{N}\sum_{n=1}^{N}\mu_n \to \mu$, the previous assumption translates into

$$\overline{X}_n \longrightarrow \mu \tag{A.53}$$

both in quadratic mean and in probability.

In reality, with sequences $\{\xi_n\}$ of stochastically independent variables, a remarkable theorem of Levy applies (see for example [10, 16]) which states the equivalence of the weak law of large numbers with the strong one, i.e., the convergence of $\overline{X}_N$ to μ occurs almost certainly, that is, with $P = 1$.

The heuristic interpretation of the CLT theorem is that in the measurement process, with a variance σ_0^2 of the error ε, accidental (i.e., with zero mean), many small, independent factors intervene with a variance smaller than σ_0^2, caused by uncontrollable causes, so that we can think that

$$\varepsilon = \sum_{n=1}^{N} \xi_n \; ; \tag{A.54}$$

assuming that the ξ_n satisfy the conditions of the Lyapunov theorem [32], it follows that ε will have the distribution (A.48).

The family of normal $\boldsymbol{X}$ regular in $\mathbb{R}^n$ is instead characterized by the probability density

$$f_{\boldsymbol{X}}(\boldsymbol{x}) = \frac{1}{(2\pi)^{1/2}(\det C)^{1/2}} e^{-\frac{1}{2}(\boldsymbol{x}-\boldsymbol{\mu})^{\top} C^{-1}(\boldsymbol{x}-\boldsymbol{\mu})} \, . \tag{A.55}$$

It is shown that $\boldsymbol{\mu}$ is the mean of $\boldsymbol{X}$,

$$\boldsymbol{\mu} = E\{\boldsymbol{X}\} \tag{A.56}$$

and that C is the covariance of $\boldsymbol{X}$

$$C = E\{(\boldsymbol{X} - \boldsymbol{\mu})(\boldsymbol{X} - \boldsymbol{\mu})^{\top}\} \; ; \tag{A.57}$$

therefore C must be a symmetric and strictly positive definite matrix in order for C^{-1} to exist.

Often the (A.55) is represented by the formula

$$\boldsymbol{X} \sim \mathcal{N}(\boldsymbol{\mu}, C) \, . \tag{A.58}$$

One can prove that in a normal distribution, all marginal distribution are normal too, namely

$$X_i \sim \mathcal{N}(\mu_i, C_{ii})\,. \tag{A.59}$$

Moreover, if one can split $\boldsymbol{X} = \begin{vmatrix} \boldsymbol{X}_{n-1} \\ X_n \end{vmatrix}$ and $\boldsymbol{\mu} = \begin{vmatrix} \mu_{n-1} \\ \mu_n \end{vmatrix}$, $C = \begin{vmatrix} C_{n-1} & \boldsymbol{C}_{n-1} \\ \boldsymbol{C}_{n-1}^{\top} & c_n \end{vmatrix}$, we find that

$$\boldsymbol{X}_{n-1} \sim \mathcal{N}(\mu_{n-1}, C_{n-1})\,. \tag{A.60}$$

The same happens when we compute the conditional distribution of one (or more) components of $\boldsymbol{X}$ given the others.

In the case it is

$$\boldsymbol{\mu} = 0\,,\ C = I \tag{A.61}$$

it is said that the normal is standardized and this is indicated as $\mathcal{Z}$,

$$\mathcal{Z} \sim \mathcal{N}(0, I)\,. \tag{A.62}$$

Note that the number of (distinct) parameters that characterize the $f_X(\boldsymbol{x})$ is equal to

$$p(\text{number of parameters}) = n + \frac{n(n+1)}{2}\,, \tag{A.63}$$

as it follows from the fact that C is symmetric and therefore has only $\frac{n(n+1)}{2}$ distinct elements.

Appendix B
Sample Variable and Maximum Likelihood

We have defined a sample $\boldsymbol{X}$ of the r.v. X of size N as the collection of N extractions from X conducted under the same conditions. Therefore, a sample is characterized by the fact that the components of $\boldsymbol{X}$ taken by themselves are all identically distributed, like the r.v. X. This is not enough to identify the distribution of the vector $\boldsymbol{X}$, called *sample variable*, however if we specify that the extractions are independent of each other, which we translate into the statement that the $\{X_i\}$ are stochastically independent, then it follows, using the (A.15)

$$f(\boldsymbol{x}) = \prod_{i=1}^{N} f_{X_i}(x_i) = \prod_{i=1}^{N} f_X(x_i)\ . \tag{B.1}$$

A sample that satisfies the (B.1) is called Bernoulli. So a sample variable has the sense of transforming a variable X with distribution $f_X(x)$ of which N values are known by extraction into an N-dimensional variable with distribution $f_{\boldsymbol{X}}(\boldsymbol{x})$ (which is given by (B.1) if the sample is Bernoullian) of which one extraction is known.

The concept then easily generalizes to the variable $\boldsymbol{X}$ of the "observables" of a system; in general, the components of $\boldsymbol{X}$ in this case are not stochastically independent, nor equally distributed, but the overall stochastic model that governs them, expressed by the density $f(\boldsymbol{x})$, is known.

It should be noted that usually the distribution of the sample variable, or the variable of the observables, is assigned only by asserting the membership of $f(\boldsymbol{x})$ to a family that also depends on a parameter vector $\boldsymbol{\vartheta} = (\vartheta_1 \dots \vartheta_p)$ to be estimated, based on the known values of the extraction $\boldsymbol{x}_0$; this is the estimation operation discussed in Chap. 3.

It is customary in the statistical literature to call the distribution model $f(\boldsymbol{x})$ likelihood function or likelihood $L(\boldsymbol{x}|\boldsymbol{\vartheta})$

$$f(\boldsymbol{x}) \equiv L(\boldsymbol{x}|\boldsymbol{\vartheta})\ . \tag{B.2}$$

F. Sansò, A. Albertella, *The Probabilistic Vision of the Physical World*, Lecture Notes in Geosystems Mathematics and Computing,
https://doi.org/10.1007/978-3-031-88268-5

Two examples are useful to clarify the concepts presented.

Example B.1 Bernoulli normal sample. In this case the underlying one-dimensional variable $f(x)$ is a normal characterized by mean μ and variance σ^2, or, given $\boldsymbol{\vartheta} = (\mu, \sigma^2)$

$$f(x|\boldsymbol{\vartheta}) = \frac{1}{\sqrt{2\pi}\sigma} e^{-\frac{1}{2}\frac{(x-\mu)^2}{\sigma^2}} \sim \mathcal{N}(\mu, \sigma^2) . \tag{B.3}$$

Applying (B.1) then we find for the sample variable $\boldsymbol{X}$

$$L(\boldsymbol{x}|\boldsymbol{\vartheta}) = \frac{1}{(2\pi)^{N/2}\sigma^N} e^{-\frac{1}{2}\sum_{i=1}^{N}\frac{(x_i-\mu)^2}{\sigma^2}} . \tag{B.4}$$

Example B.2 (Linear Model for a Normal Variable X of the Observables, Normal in $\mathbb{R}^m$) A normal variable in $\mathbb{R}^m$ has the probability density given by (A.55), however it is common to encounter a system of linear observation equations (or linearized)

$$\boldsymbol{X} = A\boldsymbol{\lambda} + \boldsymbol{a} + \boldsymbol{\varepsilon} \tag{B.5}$$

in which $\boldsymbol{\varepsilon}$ is a normal vector of accidental errors (i.e., with zero mean) and the mean of $\boldsymbol{X}$ is therefore

$$E\{\boldsymbol{X}\} = A\boldsymbol{\lambda} + \boldsymbol{a} ;$$

if this is a function of a number $n < m$ of parameters $\boldsymbol{\lambda}$ from the point of view of the estimation problem, it is an overdetermined system. Assuming that C is the covariance matrix of $\boldsymbol{\varepsilon}$, it will be

$$\boldsymbol{X} \sim \mathcal{N}(A\boldsymbol{\lambda} + \boldsymbol{a} , C) \tag{B.6}$$

or

$$L(\boldsymbol{x}|\boldsymbol{\vartheta}) = f_X(\boldsymbol{x}) = \frac{1}{(2\pi)^{N/2}\sqrt{\det C}} e^{-\frac{1}{2}(\boldsymbol{x}-A\boldsymbol{\lambda}-\boldsymbol{a})^\top C^{-1}(\boldsymbol{x}-A\boldsymbol{\lambda}-\boldsymbol{a})} \tag{B.7}$$

So in this case the parameter vector $\boldsymbol{\vartheta}$ on which L depends is $(\boldsymbol{\lambda}, C)$ and therefore has dimension $p = n + \frac{m(m+1)}{2}$.

The principle of maximum likelihood states that an "estimator" $\widehat{\vartheta}$ of ϑ is chosen in such a way that $L(\boldsymbol{X}, \widehat{\boldsymbol{\vartheta}})$ is maximum; in other words, given the deterministic function $L(\boldsymbol{x}, \boldsymbol{\vartheta})$ we look for $\widehat{\boldsymbol{\vartheta}}(\boldsymbol{x})$ for which $L(\boldsymbol{x}, \widehat{\boldsymbol{\vartheta}})$ is maximum, then instead of $\boldsymbol{x}$ we put the variable $\boldsymbol{X}$ and thus we construct the r.v. $\widehat{\boldsymbol{\vartheta}}(\boldsymbol{X})$, called maximum likelihood estimator.

Naturally, a necessary condition for $\widehat{\boldsymbol{\vartheta}}(\boldsymbol{x})$ to be of maximum likelihood, when L is duly regular, is

$$\partial_{\boldsymbol{\vartheta}} L(\boldsymbol{x}|\boldsymbol{\vartheta}) = 0 \Rightarrow \boldsymbol{\vartheta} = \widehat{\boldsymbol{\vartheta}}(\boldsymbol{x}) \, . \tag{B.8}$$

For reasons that we will see, instead of $L(\boldsymbol{x}|\boldsymbol{\vartheta})$ we prefer to use the function

$$\ell(\boldsymbol{x}|\boldsymbol{\vartheta}) = \log L(\boldsymbol{x}|\boldsymbol{\vartheta}) \, ; \tag{B.9}$$

since the logarithm is a monotonically increasing function, (B.8) becomes

$$\ell_{\boldsymbol{\vartheta}}(\boldsymbol{x}|\widehat{\boldsymbol{\vartheta}}) = 0 \, . \tag{B.10}$$

Since $\ell_{\boldsymbol{\vartheta}}$ plays an important role in the theory of maximum likelihood, in the literature the name of score function is used,

$$U(\boldsymbol{x}|\boldsymbol{\vartheta}) = \frac{\partial}{\partial \boldsymbol{\vartheta}} \log L(\boldsymbol{x}|\boldsymbol{\vartheta}) \equiv \ell_{\boldsymbol{\vartheta}}(\boldsymbol{x}|\boldsymbol{\vartheta}) \, , \tag{B.11}$$

from which discends the identity

$$U(\boldsymbol{x}|\widehat{\boldsymbol{\vartheta}}) \equiv 0 \tag{B.12}$$

A first important property of U is that the r.v. $U(\boldsymbol{X}|\boldsymbol{\vartheta})$ always has zero mean; in fact

$$E\{U(\boldsymbol{X}|\boldsymbol{\vartheta})\} = \int \frac{L_{\boldsymbol{\vartheta}}(\boldsymbol{x}|\boldsymbol{\vartheta})}{L(\boldsymbol{x}|\boldsymbol{\vartheta})} \cdot L(\boldsymbol{x}|\boldsymbol{\vartheta}) d_n x = \frac{\partial}{\partial \boldsymbol{\vartheta}} \int L(\boldsymbol{x}|\boldsymbol{\vartheta}) d_n x = 0 \tag{B.13}$$

since the integral of L over all $\mathbb{R}^n$ is identically equal to 1. Moving $\partial_{\boldsymbol{\vartheta}}$ outside the integral requires regularity conditions of $L(\boldsymbol{x}|\boldsymbol{\vartheta})$ that we assume here to be satisfied.

For simplicity of writing, considering the case where $\vartheta \in \mathbb{R}^1$, let's consider the identity

$$U_\vartheta = \frac{L_{\vartheta\vartheta}}{L} - \frac{L_\vartheta^2}{L^2} = \frac{L_{\vartheta\vartheta}}{L} - U^2 \, . \tag{B.14}$$

Multiplying Eq. (B.14) by $L(\boldsymbol{x}|\vartheta)$ and integrating over $\mathbb{R}^n$, we find

$$E\{U_\vartheta\} = \int L_{\vartheta\vartheta}(\boldsymbol{x}|\vartheta) d_n x - E\{U^2\} = -\sigma^2(U) \tag{B.15}$$

since $\int L_{\vartheta\vartheta} d_n x = 0$ for the same reason that Eq. (B.13) holds.

To study the asymptotic behavior of $\widehat{\vartheta}$ when $N \to +\infty$, we will use theorem (A.5) under the simplifying assumption that $\boldsymbol{X}$ represents a Bernoulli sample, i.e., with independent and identically distributed X_i components.

Under this assumption, for large N, we can use the CLT and state that

$$U(\boldsymbol{X}|\vartheta) = \sum_{i=1}^{N} \frac{\partial}{\partial \vartheta} \log f_X(X_i|\vartheta) \equiv \sum_{i=1}^{N} U_i(\vartheta) \tag{B.16}$$

will asymptotically have a normal distribution since the $U_i(\vartheta)$ are independent and identically distributed, as it happens to the X_i. Therefore, remembering that $U(\boldsymbol{X}|\vartheta)$ has zero mean (see (B.13)), it will be

$$U(\boldsymbol{X}|\vartheta) \sim \mathcal{N}(0, \sigma_U^2) \,. \tag{B.17}$$

On the other hand, using the definition of $\widehat{\vartheta}$ and linearizing $U(\boldsymbol{X}|\vartheta)$, we see that based on (B.12), the following approximate relationship holds

$$U(\boldsymbol{X}|\widehat{\vartheta}) \cong U(\boldsymbol{X}|\vartheta) + U_\vartheta(\boldsymbol{X}|\vartheta)(\widehat{\vartheta} - \vartheta) \,. \tag{B.18}$$

If we further approximate $U_\vartheta(\boldsymbol{X}|\vartheta)$ with its average value, since this function multiplies the increment $\widehat{\vartheta} - \vartheta$ which is assumed to be small with high probability, with (B.15) we find

$$U(\boldsymbol{X}|\vartheta) - \sigma_U^2(\widehat{\vartheta} - \vartheta) \cong 0 \,. \tag{B.19}$$

Therefore, albeit in an approximate form, we find the asymptotic relationship

$$\widehat{\vartheta} - \vartheta = \frac{U(\boldsymbol{X}|\vartheta)}{\sigma_U^2} \tag{B.20}$$

which, for (B.17), gives

$$\widehat{\vartheta} \sim \mathcal{N}\left(\vartheta, \frac{1}{\sigma_U^2}\right) \,. \tag{B.21}$$

Note that, since $U(\boldsymbol{X}|\boldsymbol{\vartheta})$ is the sum of identically and independently distributed variables, $\sigma_U^2 = O(N)$. Therefore, taking the variance of (B.20), we see that $\sigma^2(\widehat{\vartheta}) = O\left(\frac{1}{N}\right)$, demonstrating the consistency of the estimator $\widehat{\vartheta}$. On the other hand, the estimator $\widehat{\vartheta}$ is not unbiased, except asymptotically, i.e., in general

$$E\{\widehat{\vartheta}\} \neq \vartheta \tag{B.22}$$

but

$$\lim_{N\to\infty} E\{\widehat{\vartheta}\} = \vartheta \tag{B.23}$$

as seen from (B.21).

To demonstrate (B.22) a useful counterexample is enough.

Example B.3 Let $\boldsymbol{X}$ be the Bernoulli sample variable of a normal, i.e.,

$$X_i \sim \mathcal{N}(\mu, \sigma^2) \; ; \tag{B.24}$$

in this case $\boldsymbol{\vartheta} = (\mu, \sigma^2)$.

The likelihood of $\boldsymbol{X}$ is given by (B.4), so that

$$\log L(\boldsymbol{X}|\boldsymbol{\vartheta}) = -\frac{N}{2}\log 2\pi - \frac{N}{2}\log \sigma^2 - \frac{1}{2}\sum_i \frac{(X_i-\mu)^2}{\sigma^2} \,. \tag{B.25}$$

Then the maximum likelihood estimators of μ and σ^2 are the solutions of the system

$$\begin{cases} \dfrac{\sum_{i=1}^N (X_i - \widehat{\mu})^2}{\widehat{\sigma}^2} = 0 \\ -\dfrac{N}{2}\dfrac{1}{\widehat{\sigma}^2} + \dfrac{1}{2}\dfrac{\sum_{i=1}^N (X_i - \widehat{\mu})^2}{\widehat{\sigma}^4} = 0 \end{cases} \tag{B.26}$$

or

$$\begin{cases} \widehat{\mu} = \dfrac{1}{N}\displaystyle\sum_{i=1}^N X_i \\ \widehat{\sigma}^2 = \dfrac{1}{N}\displaystyle\sum_{i=1}^N (X_i - \widehat{\mu})^2 \,, \end{cases} \tag{B.27}$$

that is, sample mean and variance.

Therefore, $\widehat{\mu}$ is an unbiased estimator while $\widehat{\sigma}^2$ is not; in fact, as is known, a correct estimator of σ^2 is given by

$$\widehat{\sigma}^2_{unb} = \frac{1}{N-1}\sum_{i=1}^N (X_i - \widehat{\mu})^2 \,. \tag{B.28}$$

Appendix C
Gradient-Based Optimization

In Chap. 6 there are two optimization problems that correspond to the search for the minimum of (6.12), (6.13) with respect to $\boldsymbol{\vartheta}$ for each given λ.

In any case, the problem is always to minimize a function $F(\boldsymbol{\xi})$ of a vector $\boldsymbol{\xi}$, which we assume to have dimension d. The topic has an extensive bibliography and is still the subject of research, stimulated also by the development and diffusion of machine learning algorithms. We clearly do not aim to cover such a vast area, but we limit ourselves to a discussion on the numerical complexity of the search for the minimum, indicating, for the applications of our interest, the choices that push towards solutions making the implementation of machine learning, in particular deep machine learning, numerically feasible.

As already mentioned in the text, here we limit ourselves to considering the search for the relative minimum of $F(\boldsymbol{\xi})$ in a zone A of the space $\mathbb{R}^d$ where F has a convex behavior and where the function is continuous and bounded with its first and second derivatives. For simplicity, we can assume that the set A is itself convex, for example a sphere or a hypercube in $\mathbb{R}^d$. The idea is to start from a point $\boldsymbol{\xi}_0$ in A and look for a sequence of points $\boldsymbol{\xi}_n$ such that between $\boldsymbol{\xi}_n$ and $\boldsymbol{\xi}_{n+1}$ the $F(\boldsymbol{\xi})$ is always decreasing. In this case, there will be a limit

$$\overline{F} = \lim F(\boldsymbol{\xi}_n) \tag{C.1}$$

and if F is strictly convex, it can only be

$$\overline{F} = \min_A F(\boldsymbol{\xi}) = F(\overline{\boldsymbol{\xi}}) \tag{C.2}$$

with

$$\overline{\boldsymbol{\xi}} = \lim \boldsymbol{\xi}_n\,, \tag{C.3}$$

the only point of minimum of F.

F. Sansò, A. Albertella, *The Probabilistic Vision of the Physical World*, Lecture Notes in Geosystems Mathematics and Computing,
https://doi.org/10.1007/978-3-031-88268-5

The point is, of course, that one would like a sequence $\{\boldsymbol{\xi}_n\}$ that approximates $\overline{\boldsymbol{\xi}}$ as quickly as possible, i.e., with the smallest number of iterations.

The most commonly used algorithm for this purpose has the form

$$\boldsymbol{\xi}_{n+1} = \boldsymbol{\xi}_n - \gamma_n \nabla_{\boldsymbol{\xi}} F(\boldsymbol{\xi}_n) , \tag{C.4}$$

with different choices for the coefficient γ_n. The meaning of (C.4) is to move from $\boldsymbol{\xi}_n$ to a new point $\boldsymbol{\xi}_{n+1}$ in the opposite direction to ∇F, i.e., in the direction of maximum slope, so algorithms of this type are called "steepest descent". Other choices that improve the efficiency of convergence are possible, but here we limit ourselves to (C.4) which is the most widely used.

The choice of γ_n essentially leads to two possible algorithms, which are distinguished because one requires the calculation of the Hessian of F, i.e., of

$$D^2 F(\boldsymbol{\xi}) \equiv \nabla_{\boldsymbol{\xi}} \nabla_{\boldsymbol{\xi}}^{\top} F(\boldsymbol{\xi}) \equiv H(\boldsymbol{\xi}) \tag{C.5}$$

and is therefore called second order, while the other fixes the decrease rate of γ_n a priori, when $n \to \infty$, and therefore requires only the knowledge of the gradient of F,

$$\boldsymbol{h}(\boldsymbol{\xi}) = \nabla_{\boldsymbol{\xi}} F(\boldsymbol{\xi}) ; \tag{C.6}$$

therefore the method is called first order, or, in the literature on machine learning, "learning by gradient" [3].

The first method, which is a generalization to $\mathbb{R}^d$ of the classic Newton–Raphson method, consists in choosing γ_n according to the rule

$$\gamma_n = \arg\min_{\gamma} F[\boldsymbol{\xi}_n - \gamma \boldsymbol{h}(\boldsymbol{\xi}_n)] . \tag{C.7}$$

We put for simplicity

$$\boldsymbol{h}_n = \boldsymbol{h}(\boldsymbol{\xi}_n) , \; H_n = H(\boldsymbol{\xi}_n) . \tag{C.8}$$

The meaning of (C.7) is to move along the line $\boldsymbol{\xi}_n - \gamma \boldsymbol{h}_n$, i.e., in the direction of maximum slope at $\boldsymbol{\xi}_n$, until finding the minimum of $F(\boldsymbol{\xi}_n - \gamma \boldsymbol{h}_n)$.

From the minimum condition (C.7) we find for γ the equation, generally nonlinear,

$$\frac{d}{d\gamma} F(\boldsymbol{\xi}_n - \gamma \boldsymbol{h}_n) = -\boldsymbol{h}^{\top}(\boldsymbol{\xi}_n - \gamma \boldsymbol{h}_n)\boldsymbol{h}_n = 0 . \tag{C.9}$$

Linearizing $\boldsymbol{h}^{\top}$ with respect to $\gamma \boldsymbol{h}_n$ we find

$$[-\boldsymbol{h}_n + \gamma H_n \boldsymbol{h}_n]^{\top} \boldsymbol{h}_n = 0 \tag{C.10}$$

from which

$$\gamma_n = \frac{\boldsymbol{h}_n^\top \boldsymbol{h}_n}{\boldsymbol{h}_n^\top H_n \boldsymbol{h}_n} \; ; \tag{C.11}$$

we recall that the Hessian of a regular function F is always a symmetric matrix, and in our case, assuming F strictly convex, it is also positive definite.

As can be seen, the calculation of (C.11), to then return to (C.4), requires the knowledge of the second derivatives. We note that in the case of our interest the $F(\boldsymbol{\xi})$ has the form

$$F(\boldsymbol{\xi}) = \frac{1}{N}\sum_{i=1}^{N} F_i(\boldsymbol{\xi}) \, , \tag{C.12}$$

where N is on the order of the sample size and can easily have, for example, the value $N = 10^6$. Therefore, even assuming we know the analytical form of $\boldsymbol{h}(\boldsymbol{\xi})$ and $H(\boldsymbol{\xi})$, which is not always easy, to obtain F, $\boldsymbol{h}$ or H at a point $\boldsymbol{\xi}$ we must calculate:

$$\begin{array}{lll} F(\boldsymbol{\xi}) & N & \text{functions} \\ \boldsymbol{h}(\boldsymbol{\xi}) & N \cdot d & \text{functions} \\ H(\boldsymbol{\xi}) & N \cdot d^2 & \text{functions.} \end{array}$$

With a number of parameters $d \sim 10^2$ or even more, it is clear that the calculation of $F, \boldsymbol{h}, H$, even for just one point, quickly becomes prohibitive. Therefore, a prudential decision is to give up on second-order optimization and only use first-order algorithms.

The problem of choosing the sequence $\{\gamma_n\}$ remains.

For this purpose, we formulate a simple assumption, that is, we assume that $H(\boldsymbol{\xi})$, which as mentioned must be symmetric and positive definite, satisfies the condition

$$\forall \boldsymbol{u} \in \mathbb{R}^d \, , \; a|\boldsymbol{u}|^2 \le \boldsymbol{u}^\top H(\boldsymbol{\xi})\boldsymbol{u} \le b|\boldsymbol{u}|^2 \, , \tag{C.13}$$

throughout the area A where we are looking for the minimum. Essentially, (C.13) states that the minimum eigenvalue of H is lower bounded in A and the maximum eigenvalue is upper bounded. Based on this observation, we also note that it must be

$$\forall \boldsymbol{u} \in \mathbb{R}^d \, , \; |H\boldsymbol{u}| = \sqrt{\boldsymbol{u}^\top H^2 \boldsymbol{u}} \le b|\boldsymbol{u}| \tag{C.14}$$

and similarly

$$\forall \boldsymbol{u} \in \mathbb{R}^d \; |H\boldsymbol{u}| \ge a|\boldsymbol{u}| \, . \tag{C.15}$$

Furthermore, we can derive the relation, which will be useful shortly,

$$\forall \boldsymbol{\xi}', \boldsymbol{\xi} \in A\,,\; a|\boldsymbol{\xi}' - \boldsymbol{\xi}| \leq |\boldsymbol{h}(\boldsymbol{\xi}') - \boldsymbol{h}(\boldsymbol{\xi})| \leq b|\boldsymbol{\xi}' - \boldsymbol{\xi}|\,; \tag{C.16}$$

this comes from an application to (C.13) , namely the Taylor formula

$$\boldsymbol{h}(\boldsymbol{\xi}') - \boldsymbol{h}(\boldsymbol{\xi}) = H(\boldsymbol{\xi} + \tilde{t}(\boldsymbol{\xi}' - \boldsymbol{\xi}))(\boldsymbol{\xi}' - \boldsymbol{\xi})\,, \tag{C.17}$$

valid for some $\tilde{t}$ between 0 and 1, and from the observation that, being $\boldsymbol{\xi}, \boldsymbol{\xi}' \in A$, it will also be $\boldsymbol{\xi} + \tilde{t}(\boldsymbol{\xi}' - \boldsymbol{\xi}) \in A$, due to the supposed convexity of A.

We are now in a position to prove the following convergence theorem.

Theorem C.1 *A sufficient condition for the convergence of the sequence*

$$\boldsymbol{\xi}_{n+1} = \boldsymbol{\xi}_n - \gamma_n \boldsymbol{h}_n \tag{C.18}$$

to the minimum point, $\bar{\boldsymbol{\xi}}$, of a F convex in A, satisfying condition (C.13), is that

$$\gamma_n < \frac{1}{b} \tag{C.19}$$

and that

$$\sum_{n=0}^{+\infty} \gamma_n = +\infty\,. \tag{C.20}$$

In fact, noting that $\bar{\boldsymbol{h}} = \boldsymbol{h}(\bar{\boldsymbol{\xi}}) = 0$ because $\bar{\boldsymbol{\xi}}$ is a minimum point, using (C.18) and (C.16) we can write

$$\begin{aligned}|\boldsymbol{\xi}_{n+1} - \bar{\boldsymbol{\xi}}| = |\boldsymbol{\xi}_n - \bar{\boldsymbol{\xi}} - \gamma_n \boldsymbol{h}_n| = |\boldsymbol{\xi}_n - \bar{\boldsymbol{\xi}} - \gamma_n(\boldsymbol{h}_n - \bar{\boldsymbol{h}})| \geq \\ \geq |\boldsymbol{\xi}_n - \bar{\boldsymbol{\xi}}| - \gamma_n|\boldsymbol{h}_n - \bar{\boldsymbol{h}}| \geq |\boldsymbol{\xi}_n - \bar{\boldsymbol{\xi}}|(1 - \gamma_n b)\end{aligned} \tag{C.21}$$

Therefore

$$|\boldsymbol{\xi}_{n+1} - \bar{\boldsymbol{\xi}}| \leq \rho_n |\boldsymbol{\xi}_0 - \bar{\boldsymbol{\xi}}| \tag{C.22}$$

with

$$\rho_n = \prod_{k=0}^{n} (1 - \gamma_k b)\,. \tag{C.23}$$

But then, using the inequality

$$\log(1 - x) \leq -x \quad (0 < x < 1)\,, \tag{C.24}$$

we can write

$$\rho_n = e^{\sum_{k=0}^{n} \log(1-\gamma_k b)} < e^{-b\sum_{k=0}^{n} \gamma_k}\ , \tag{C.25}$$

so that $\rho_n \to 0$ and $\boldsymbol{\xi}_{n+1} \to \bar{\boldsymbol{\xi}}$ if (C.20) is satisfied.

We note that conditions (C.19), (C.20) can also be satisfied by a constant sequence $\{\gamma_n\}$, $\gamma_n = \gamma$, provided γ is sufficiently small. However, an argument that we will see shortly, allowing to significantly reduce the complexity of the calculation of $\boldsymbol{h}(\boldsymbol{\xi})$, suggests that the $\{\gamma_n\}$ should at least satisfy the relation

$$\sum_{n=0}^{+\infty} \gamma_n^2 < +\infty\ . \tag{C.26}$$

The conclusion is that a sequence of the type

$$\gamma_n = \frac{c}{1+n} \tag{C.27}$$

seems to be satisfactory, adapting the constant c to the specific problem.

We now come to discuss a way to significantly limit the complexity of the calculation of

$$\boldsymbol{h}(\boldsymbol{\xi}) = \nabla F(\boldsymbol{\xi}) = \frac{1}{N}\sum_{i=1}^{N} \nabla F_i(\boldsymbol{\xi})\ . \tag{C.28}$$

The method is called *stochastic gradient*, and it consists in remembering that the $F_i(\boldsymbol{\xi})$ to be used in (C.12) to define the F to be minimized, in our case, derive from a single function, $g(\boldsymbol{X}|\boldsymbol{\vartheta})$ or $Q(Y, \boldsymbol{X}|\boldsymbol{\vartheta})$, calculated for the values $(Y_i, \boldsymbol{X}_i)$, a very large sample drawn from the r.v. $(Y, \boldsymbol{X})$. Naturally this means not considering the term $\lambda|\boldsymbol{\vartheta}|^2$, which however does not present calculation difficulties and which we therefore ignore for the rest of this appendix. So if we call $f(Y, \boldsymbol{X}|\boldsymbol{\xi})$ the function $g(\boldsymbol{X}|\boldsymbol{\xi})$ or $Q(Y, \boldsymbol{X}|\boldsymbol{\xi})$ that interests us, (C.18), with a certain abuse of notation, has the form

$$F(\boldsymbol{\xi}) = \frac{1}{N}\sum_{i=1}^{N} f(Y_i, \boldsymbol{X}_i|\boldsymbol{\xi})\ . \tag{C.29}$$

As is evident, this appears as an empirical average on the sample $\{f(Y_i, \boldsymbol{X}_i|\boldsymbol{\xi})\}$; but, having supposed that the $\{Y_i, \boldsymbol{X}_i\}$ are independent, we can randomly extract a subsample $\{Y_{i_k}, \boldsymbol{X}_{i_k}\}$, of size $M << N$, from $\{Y_i, \boldsymbol{X}_i\}$ and calculate on this the empirical average, taking

$$F(\boldsymbol{\xi}) \cong \widetilde{F}(\boldsymbol{\xi}) = \frac{1}{M}\sum_{k=1}^{M} f(Y_{i_k}, \boldsymbol{X}_{i_k}|\boldsymbol{\xi}) \tag{C.30}$$

and therefore

$$\boldsymbol{h}(\boldsymbol{\xi}) \cong \widetilde{\boldsymbol{h}}(\boldsymbol{\xi}) = \frac{1}{M}\sum_{k=1}^{M} \nabla_{\boldsymbol{\xi}} f(Y_{i_k}, \boldsymbol{X}_{i_k}|\boldsymbol{\xi})\,. \tag{C.31}$$

In this way instead of having a sample for example of 10^6 elements we can work with a "minibatch" of a few hundred elements, leaving to the central theorem of statistics to handle the approximation of (C.28) with (C.31).

Naturally the use of a minibatch greatly reduces the complexity of the calculation but introduces an error in the determination of $\overline{\boldsymbol{\xi}}$.

It is therefore necessary to verify that the variability of the error does not cause the sequence $\widetilde{\boldsymbol{\xi}}_n$ determined with the approximate gradient (C.31) to remain far from the sequence $\boldsymbol{\xi}_n$, determined with the exact gradient (C.28).

We set

$$\boldsymbol{\xi}_{n+1} = \boldsymbol{\xi}_n - \gamma_n \boldsymbol{h}_n \tag{C.32}$$

$$\widetilde{\boldsymbol{\xi}}_{n+1} = \widetilde{\boldsymbol{\xi}}_n - \gamma_n \widetilde{\boldsymbol{h}}_n \tag{C.33}$$

$$\boldsymbol{\varepsilon}_n = \boldsymbol{\xi}_n - \widetilde{\boldsymbol{\xi}}_n \tag{C.34}$$

$$\boldsymbol{\eta}_n = \boldsymbol{h}_n - \widetilde{\boldsymbol{h}}_n \tag{C.35}$$

and we note that therefore

$$\boldsymbol{\varepsilon}_{n+1} = \boldsymbol{\varepsilon}_n - \gamma_n \boldsymbol{\eta}_n\,. \tag{C.36}$$

The (C.36) implies that

$$\boldsymbol{\varepsilon}_{n+1} = -\sum_{k=0}^{n} \gamma_k \boldsymbol{\eta}_k\,, \tag{C.37}$$

considering that $\boldsymbol{\varepsilon}_0 = 0$ since both the sequences $\{\boldsymbol{\xi}_n\}$ and $\{\widetilde{\boldsymbol{\xi}}_n\}$ start from the same point $\boldsymbol{\xi}_0$.

A rigorous analysis of the error $\boldsymbol{\varepsilon}_{n+1}$ is beyond the scope of this text, however it is understood that in order for $\boldsymbol{\varepsilon}_{n+1}$ to be small it is necessary for (C.37) that at least the condition

$$\sum_{k=0}^{+\infty} \gamma_k^2 < +\infty \, .$$

This would be obvious if the $\{\eta_k\}$ were independent of each other (in any way an unrealistic assumption) considering that $\boldsymbol{h}_n$ and $\widetilde{\boldsymbol{h}}_n$ are limited quantities. Moreover, the same condition is required when extending the analysis presented here to more general cases where the $F(\boldsymbol{\xi})$ has less regularity and the minimum points of F are a set consisting not of a single point and maybe this set is not even limited. In this regard, see for example [31], Chap. 4. This justifies the choice made in (C.26).

Summarizing the discussion presented in this appendix, we conclude that the optimization problems posed are numerically implementable by choosing the gradient learning algorithm (C.18) with a sequence $\{\gamma_n\}$ of the type (C.27) and the gradient of F calculated as stochastic gradient from a minibatch (see (C.31)) randomly extracted from the total sample.

Appendix D
The Concatenated Optimization for ϑ and λ

The purpose of the appendix is to present a possible algorithm for the realization of the chain optimization, described in Sect. 6.3,

$$\vartheta(\lambda) = \arg\min_{\vartheta} F(\vartheta, \lambda) \ , \tag{D.1}$$

$$\begin{cases} \overline{\lambda} = \arg\min_{\lambda} C[\vartheta(\lambda)] \\ \overline{\vartheta} = \vartheta(\overline{\lambda}) \end{cases} \ , \tag{D.2}$$

where $F(\vartheta, \lambda)$ and $C(\lambda)$ are respectively $J_{\lambda}(\vartheta)$ (see (6.12)) and $CV(\lambda)$ (see (6.16)) for the regression problem, or $K_{\lambda}(\vartheta)$ (see (6.13)) and $C\ell(\lambda)$ (see (6.17)) for the indirect problem.

The tool at our disposal to implementing this operation is the steepest descent algorithm, described in Appendix C, realized through the use of a neural network.

However, the algorithm can provide a value of ϑ for a fixed λ, not the function $\vartheta(\lambda)$. If one wants to avoid a numerically heavy trial and error procedure, one can resort to a combination of steepest descent and linearization, as traditionally applied in least squares problems, avoiding the calculation of the second derivatives of the functions $f_i(\vartheta)$ that make up $F(\vartheta, \lambda)$ (see (6.20)).

We preliminarily recall that it is good practice to "normalize" the components of $\{\boldsymbol{X}_i\}$ as this generally increases the numerical stability of the algorithms. By this we mean that given the sample of values of the component X_k, $\{X_{k,i}\}$, and determined the maximum and minimum values

$$m_k = \min_i X_{k,i} \quad M_k = \max_i X_{k,i} \tag{D.3}$$

we define new variables

$$X'_{k,i} = \frac{X_{k,i} - m_k}{M_k - m_k} \tag{D.4}$$

F. Sansò, A. Albertella, *The Probabilistic Vision of the Physical World*, Lecture Notes in Geosystems Mathematics and Computing,
https://doi.org/10.1007/978-3-031-88268-5

which will have a range between 0 and 1. The same procedure is applied to the sample of values $\{Y_i\}$. We note that these linear transformations only involve one component at a time and therefore do not change the scheme of the neural network used for the steepest descent.

Since regression and maximum likelihood have some approximation difference, we first deal with regression, then indicating the points where the second solution differs from the first.

Clarifying the notation already used previously in Sect. 6.3 and in Appendix C, we set, for the case of regression,

$$v_i(\boldsymbol{\vartheta}) = Y_i - g(\boldsymbol{X}_i|\boldsymbol{\vartheta}) \tag{D.5}$$

so that the function $F(\boldsymbol{\vartheta}, \lambda)$ to be minimized is written

$$F(\boldsymbol{\vartheta}, \lambda) = \frac{1}{N}\sum_{i=1}^{N} v_i^2 + \lambda|\boldsymbol{\vartheta}|^2 . \tag{D.6}$$

We immediately note that here, as in the following, the index i runs along the sample, which for the $F(\boldsymbol{\vartheta}, \lambda)$ will be the training sample, hence $N = N_{Tr}$, while later for the $CV(\lambda)$ it will be the test sample, hence $N = N_{te} = N - N_{tr}$. In both cases, however, for convenience of notation, the sample can be replaced by a minibatch, i.e. $N = M$.

We also set

$$\boldsymbol{h}(\boldsymbol{\vartheta}) = \frac{1}{N}\sum_{i=1}^{N} v_i(\boldsymbol{\vartheta})\nabla_{\boldsymbol{\vartheta}}\, g(\boldsymbol{X}_i|\boldsymbol{\vartheta}) \tag{D.7}$$

so that

$$\nabla_{\boldsymbol{\vartheta}} F(\boldsymbol{\vartheta}, \lambda) = -2\boldsymbol{h}(\boldsymbol{\vartheta}) + 2\lambda\boldsymbol{\vartheta} . \tag{D.8}$$

It is clear that the function $\boldsymbol{\vartheta}(\lambda)$ of formula (D.1) is the solution of the equation

$$\boldsymbol{h}(\boldsymbol{\vartheta}) - \lambda\boldsymbol{\vartheta} = 0 . \tag{D.9}$$

We now see how to find the pair $(\boldsymbol{\vartheta}_0, \lambda_0)$ from which to start an iterative solution of the problem.

We start from an arbitrary $\boldsymbol{\vartheta}'_0$, with small but non-zero components, so that we can set

$$\lambda_0 = \frac{|\boldsymbol{h}(\boldsymbol{\vartheta}'_0)|}{|\boldsymbol{\vartheta}'_0|} . \tag{D.10}$$

It is clear that if $\boldsymbol{\vartheta}_0'$ were the solution of (D.9) with $\lambda = \lambda_0$, then (D.1) would hold, but in general this is not the case; therefore we move on to define $\boldsymbol{\vartheta}_0$ as the actual solution of the equation

$$\boldsymbol{h}(\boldsymbol{\vartheta}_0) - \lambda_0\, \boldsymbol{\vartheta}_0 = \boldsymbol{h}_0 - \lambda_0 \boldsymbol{\vartheta}_0 = 0\,, \tag{D.11}$$

obtained from the steepest descent algorithm applied to $F(\boldsymbol{\vartheta}, \lambda_0)$.

Given $(\boldsymbol{\vartheta}_0, \lambda_0)$, we need to indicate how to move to a new pair $(\boldsymbol{\vartheta}_1, \lambda_1)$ and then iterate the algorithm. Unfortunately, although we know $\boldsymbol{\vartheta}_0 = \boldsymbol{\vartheta}(\lambda_0)$, we do not know how $\boldsymbol{\vartheta}(\lambda)$ varies around λ_0 to move to the minimization of $CV(\lambda)$. Therefore we temporarily resort to the linearization of the problem, putting

$$\boldsymbol{\vartheta} = \boldsymbol{\vartheta}_0 + \boldsymbol{\eta}\,,\ \lambda = \lambda_0 + \mu\,. \tag{D.12}$$

Given $\boldsymbol{h}_0 = \boldsymbol{h}(\boldsymbol{\vartheta}_0)$ and defined the matrix

$$G_0 = \frac{1}{N}\sum_{i=1}^{N} \nabla_{\vartheta} g(\boldsymbol{X}_i|\boldsymbol{\vartheta}_0)\nabla_{\vartheta}^{\top} g(\boldsymbol{X}_i|\boldsymbol{\vartheta}_0)\,, \tag{D.13}$$

we substitute (D.12) into $F(\boldsymbol{\vartheta}, \lambda)$.

In particular we set

$$\begin{aligned} v_i^2 &\equiv [Y_i - g(\boldsymbol{X}_i|\boldsymbol{\vartheta}_0) - \nabla_{\vartheta}^{\top} g(\boldsymbol{X}_i|\boldsymbol{\vartheta}_0)\boldsymbol{\eta}]^2 \\ &\equiv [v_{i0} - \nabla_{\vartheta}^{\top} g(\boldsymbol{X}_i(\boldsymbol{\vartheta}_0)\boldsymbol{\eta}]^2\,. \end{aligned} \tag{D.14}$$

Developing $F(\boldsymbol{\vartheta}_0 + \boldsymbol{\eta}, \lambda_0 + \mu)$ up to the second order terms we get

$$\begin{aligned} F(\boldsymbol{\vartheta}_0 + \boldsymbol{\eta}, \lambda_0 + \mu) = {} & F(\boldsymbol{\vartheta}_0, \lambda_0) - 2\boldsymbol{h}_0^{\top}\boldsymbol{\eta} + \boldsymbol{\eta}^{\top} G_0 \boldsymbol{\eta} + \\ & + \lambda_0(2\boldsymbol{\vartheta}_0^{\top}\boldsymbol{\eta} + \boldsymbol{\eta}^{\top}\boldsymbol{\eta}) + \mu(|\boldsymbol{\vartheta}_0|^2 + 2\boldsymbol{\vartheta}_0^{\top}\boldsymbol{\eta})\,. \end{aligned} \tag{D.15}$$

Now the minimization with respect to $\boldsymbol{\eta}$ is standard and leads to the equation

$$-2\boldsymbol{h}_0 + 2G_0\boldsymbol{\eta} + 2\lambda_0\boldsymbol{\vartheta}_0 + 2\lambda_0\boldsymbol{\eta} + 2\mu\boldsymbol{\vartheta}_0 = 0\,; \tag{D.16}$$

recalling (D.11) this equation becomes

$$(G_0 + \lambda_0)\boldsymbol{\eta} + \mu\boldsymbol{\vartheta}_0 = 0\,, \tag{D.17}$$

which, when solved, gives

$$\boldsymbol{\eta} = -\mu(G_0 + \lambda_0)^{-1}\boldsymbol{\vartheta}_0\,. \tag{D.18}$$

This is the analytical relationship between η and μ that we were looking for and that, although approximate, we use to minimize $CV(\boldsymbol{\vartheta}(\lambda))$.

For this purpose, we calculate, by linearizing,

$$CV(\boldsymbol{\vartheta}_0 + \boldsymbol{\eta}(\mu)) \cong \frac{1}{N_{te}} \sum_{i=N_{tr}+1}^{N} [Y_i - g(\boldsymbol{X}_i|\boldsymbol{\vartheta}_0) - \nabla_{\vartheta}^{\top} g(\boldsymbol{X}_i|\boldsymbol{\vartheta}_0)\boldsymbol{\eta}]^2 ; \quad \text{(D.19)}$$

we note that in (D.19) the index i runs on the test sample, or on a minibatch extracted from this.

Let

$$v_{i0} = Y_i - g(\boldsymbol{X}_i|\boldsymbol{\vartheta}_0) , \quad \text{(D.20)}$$

$$\boldsymbol{h}_{te} = \frac{1}{N_{te}} \sum_{i=N_{tr}+1}^{N} v_{i0} \nabla_{\vartheta} g(\boldsymbol{X}_i|\boldsymbol{\vartheta}_0) , \quad \text{(D.21)}$$

$$G_{te} = \frac{1}{N_{te}} \sum_{i=N_{tr}+1}^{N} \nabla_{\vartheta} g(\boldsymbol{X}_i|\boldsymbol{\vartheta}_0) \nabla_{\vartheta}^{\top} g(\boldsymbol{X}_i|\boldsymbol{\vartheta}_0) , \quad \text{(D.22)}$$

we find

$$CV(\boldsymbol{\vartheta}_0 + \boldsymbol{\eta}(\mu)) \cong CV(\boldsymbol{\vartheta}_0) - 2\boldsymbol{h}_{te}^{\top}\boldsymbol{\eta} + \boldsymbol{\eta}^{\top} G_{te} \boldsymbol{\eta} . \quad \text{(D.23)}$$

The minimum condition with respect to μ will therefore be

$$-2\boldsymbol{h}_{te}^{\top}\dot{\eta} + 2\eta^{\top} G_{te}\dot{\eta} = 0 \quad \text{(D.24)}$$

where we have indicated

$$\dot{\eta} = \frac{d\eta}{d\mu} = -(G_0 + \lambda_0)^{-1}\boldsymbol{\vartheta}_0 . \quad \text{(D.25)}$$

By using (D.18) and (D.25) we see that (D.24) is a linear equation in μ, which when solved gives

$$\mu_1 = -\frac{\boldsymbol{h}_{te}^{\top}(G_0 + \lambda_0)^{-1}\boldsymbol{\vartheta}_0}{\boldsymbol{\vartheta}_0^{\top}(G_0 + \lambda_0)^{-1} G_{te} (G_0 + \lambda_0)^{-1}\boldsymbol{\vartheta}_0} . \quad \text{(D.26)}$$

From μ_1 we can now derive

$$\begin{cases} \lambda_1 = \lambda_0 + \mu_1 , \\ \eta_1 = -\mu_1 (G_0 + \lambda_0)^{-1}\boldsymbol{\vartheta}_0 , \\ \boldsymbol{\vartheta}_1' = \boldsymbol{\vartheta}_0 + \boldsymbol{\eta}_1 . \end{cases} \quad \text{(D.27)}$$

We take λ_1 as a step forward from λ_0 in the sequence of $\{\lambda_n\}$, and define a new $\boldsymbol{\vartheta}_1$ as

$$\boldsymbol{\vartheta}_1 = \arg\min_{\boldsymbol{\vartheta}} F(\boldsymbol{\vartheta}, \lambda_1)\ , \tag{D.28}$$

obtained by steepest descent, starting from the approximate value $\boldsymbol{\vartheta}'_1$.

Having obtained $(\boldsymbol{\vartheta}_1, \lambda_1)$ we can repeat the algorithm constructing the sequence $(\boldsymbol{\vartheta}_n, \lambda_n)$.

An analysis of the convergence of the method depends on the form of the family $g(\boldsymbol{X}|\boldsymbol{\vartheta})$ and cannot be conducted in general. However, it is seen that if $(\boldsymbol{\vartheta}_n, \lambda_n)$ is convergent, that is if

$$\mu_n \to 0\ ,\ \eta_n \to 0\ \lambda_n \to \overline{\lambda}\ , \boldsymbol{\vartheta}_n \to \overline{\boldsymbol{\vartheta}} \tag{D.29}$$

then $\overline{\lambda}$, $\overline{\boldsymbol{\vartheta}}$ are solutions of the stationarity equations

$$\begin{cases} \boldsymbol{h}(\overline{\boldsymbol{\vartheta}}) - \overline{\lambda}\overline{\boldsymbol{\vartheta}} = 0 \\ \\ \frac{d}{d\lambda} CV(\boldsymbol{\vartheta}(\lambda))\big|_{\lambda=\overline{\lambda}} = 0 \end{cases} \tag{D.30}$$

which correspond to the necessary minimum conditions of (D.1) and (D.2).

We now move on to the case of maximum likelihood. In this case it will be (see (6.13))

$$K_\lambda(\boldsymbol{\vartheta}) = -\frac{1}{N}\sum_{i=1}^{N} \log Q(Y_i, \boldsymbol{X}_i|\boldsymbol{\vartheta}) + \lambda|\boldsymbol{\vartheta}|^2\ ; \tag{D.31}$$

as usual we use for brevity the notation $Q_i(\boldsymbol{\vartheta}) = Q(Y_i, \boldsymbol{X}_i)$.

Recalling the notation of Appendix B, we will have

$$\boldsymbol{h}(\boldsymbol{\vartheta}) = \frac{1}{N}\sum_{i=1}^{N} \frac{\nabla_\vartheta Q_i(\boldsymbol{\vartheta})}{Q_i(\boldsymbol{\vartheta})} = \frac{1}{N}\sum_{i=1}^{N} \boldsymbol{U}_i(\boldsymbol{\vartheta}) \tag{D.32}$$

and the stationarity equation that determines $\boldsymbol{\vartheta}(\lambda)$ becomes

$$\boldsymbol{h}(\boldsymbol{\vartheta}) - 2\lambda\boldsymbol{\vartheta} = 0\ . \tag{D.33}$$

Therefore, we can trigger the iterative algorithm by choosing $\boldsymbol{\vartheta}'_0 \neq 0$ and setting

$$\lambda_0 = \frac{|\boldsymbol{h}(\boldsymbol{\vartheta}'_0)|}{2|\boldsymbol{\vartheta}'_0|}\ ; \tag{D.34}$$

subsequently we determine $\boldsymbol{\vartheta}_0$ from

$$\boldsymbol{\vartheta}_0 = \arg\min_{\boldsymbol{\vartheta}} K_\lambda(\boldsymbol{\vartheta}, \lambda_0) \tag{D.35}$$

with the steepest descent algorithm.

We note that $\boldsymbol{\vartheta}_0$ will then satisfy the equation

$$\boldsymbol{h}(\boldsymbol{\vartheta}_0) - 2\lambda_0 \boldsymbol{\vartheta}_0 = 0\,. \tag{D.36}$$

To study the behavior of $\boldsymbol{\vartheta}(\lambda)$ around $\boldsymbol{\vartheta}_0 = \boldsymbol{\vartheta}(\lambda_0)$, we proceed directly to linearize Eq. (D.33) by setting

$$\lambda_1 = \lambda_0 + \mu\,,\ \boldsymbol{\vartheta}_1 = \boldsymbol{\vartheta}_0 + \boldsymbol{\eta}\,. \tag{D.37}$$

This naturally requires the calculation of the first derivatives of $\boldsymbol{U}_i$ with respect to $\boldsymbol{\vartheta}$, a point that differentiates this analysis from that of regression. We will return to this point later.

We set

$$I(\boldsymbol{\vartheta}) = -\nabla_\vartheta \boldsymbol{h}^\top(\boldsymbol{\vartheta}) = -\frac{1}{N}\sum_{i=1}^{N} \nabla_\vartheta \boldsymbol{U}_i^\top(\boldsymbol{\vartheta}) \tag{D.38}$$

and for simplicity

$$I_0 = I(\boldsymbol{\vartheta}_0)\,,\ \boldsymbol{h}_0 = \boldsymbol{h}(\boldsymbol{\vartheta}_0)\,. \tag{D.39}$$

The equation $\nabla_\vartheta K_\lambda(\boldsymbol{\vartheta}) = 0$ when linearized gives us

$$-\,I_0\boldsymbol{\eta} - 2\mu\boldsymbol{\vartheta}_0 - 2\lambda_0\boldsymbol{\eta} = 0\,, \tag{D.40}$$

which, when solved, becomes

$$\boldsymbol{\eta} = -2\mu(I_0 + 2\lambda_0)^{-1}\boldsymbol{\vartheta}_0\,, \tag{D.41}$$

with an obvious analogy with Eq. (D.18).

We note that, as in Eq. (D.24),

$$\frac{d}{d\mu}\boldsymbol{\eta}(\mu) = \dot{\boldsymbol{\eta}}(\mu) = -2(I_0 + 2\lambda_0)^{-1}\boldsymbol{\vartheta}_0\,. \tag{D.42}$$

To optimize μ, we return to the definition of cross-likelihood

$$C\ell(\vartheta) = -\frac{1}{N_{te}}\sum_{N_{tr}+1}^{N} \log Q_i(\boldsymbol{\vartheta})\,, \tag{D.43}$$

noting that now the index i runs on the test sample or on its minibatch. To optimize the cross-likelihood with respect to μ, we develop it up to the second order in $\boldsymbol{\eta}$, setting

$$\begin{cases} \boldsymbol{h}_{te}(\boldsymbol{\vartheta}) = \frac{1}{N_{te}} \sum_{i=N_{tr}+1}^{N} \boldsymbol{U}_i \\ I_{te}(\boldsymbol{\vartheta}) = -\nabla_{\boldsymbol{\vartheta}} \boldsymbol{h}_{te}^{\top}(\boldsymbol{\vartheta})\ . \end{cases} \tag{D.44}$$

We thus have

$$C\ell(\boldsymbol{\vartheta}_0 + \boldsymbol{\eta}) = C\ell(\boldsymbol{\vartheta}_0) - \boldsymbol{h}_{te}(\boldsymbol{\vartheta}_0) + \frac{1}{2}\boldsymbol{\eta}^{\top} I_{te}(\vartheta_0)\boldsymbol{\eta} \tag{D.45}$$

where $\boldsymbol{\eta}$ is given by Eq. (D.41).

Imposing the derivative to be zero

$$\frac{d}{d\mu} C\ell(\boldsymbol{\vartheta}_0 + \boldsymbol{\eta}) = (-h_{te}(\boldsymbol{\vartheta}_0) + I_{te}(\boldsymbol{\vartheta}_0)\boldsymbol{\eta})^{\top}\dot{\boldsymbol{\eta}} = 0 \tag{D.46}$$

we find for μ the value

$$\mu_1 = -\frac{\boldsymbol{h}_{te}^{\top}(\boldsymbol{\vartheta}_0)(I_0 + 2\lambda_0)^{-1}\boldsymbol{\vartheta}_0}{2\boldsymbol{\vartheta}_0^{\top}(I_0 + 2\lambda_0)^{-1} I_{te}(\boldsymbol{\vartheta}_0)(I_0 + 2\lambda_0)^{-1}\boldsymbol{\vartheta}_0}\ . \tag{D.47}$$

We now have

$$\lambda_1 = \lambda_0 + \lambda_1\ ,\ \boldsymbol{\vartheta}_1' = \boldsymbol{\vartheta}_0 + \boldsymbol{\eta}(\mu_1) \tag{D.48}$$

and so we can proceed to determine

$$\boldsymbol{\vartheta}_1 = \arg\min_{\boldsymbol{\vartheta}} K_{\lambda_1}(\boldsymbol{\vartheta})\ . \tag{D.49}$$

The transition $(\boldsymbol{\vartheta}_0, \lambda_0) \to (\boldsymbol{\vartheta}_1, \lambda_1)$ can now be iterated.

As already observed, the procedure designed in this way requires the calculation of the second derivatives of $Q(Y, \boldsymbol{X}|\boldsymbol{\vartheta})$ with respect to $\boldsymbol{\vartheta}$. However, with a further approximation, this can be avoided based on the following reasoning. First, we observe that

$$I(\boldsymbol{\vartheta}) = \frac{1}{N}\sum_{i=1}^{N} \nabla_{\boldsymbol{\vartheta}} \boldsymbol{U}_i^{\top} \cong -\nabla_{\boldsymbol{\vartheta}} E_P\{\boldsymbol{U}^{\top}(\boldsymbol{\vartheta})\} = \\ = -\int \nabla_{\boldsymbol{\vartheta}} \boldsymbol{U}^{\top}(\vartheta) P(Y, \boldsymbol{X}) dy dx\ . \tag{D.50}$$

But

$$\nabla_{\vartheta} U^{\top}(\vartheta) = -\frac{\nabla_{\vartheta} Q(\vartheta) \nabla_{\vartheta}^{\top} Q(\vartheta)}{Q^2(\vartheta)} + \frac{\nabla_{\vartheta} \nabla_{\vartheta}^{\top} Q}{Q(\vartheta)} \,. \tag{D.51}$$

If we can assume that

$$Q(Y; X|\vartheta) \sim P(Y, X) \,, \tag{D.52}$$

using Eq. (D.51) in Eq. (D.50) and noting that

$$\int \nabla_{\vartheta} \nabla_{\vartheta}^{\top} Q(\vartheta) dy dX \equiv 0 \tag{D.53}$$

we find

$$\begin{aligned} I(\vartheta) &\cong \int U(\vartheta) U^{\top}(\vartheta) P(y, X) dy dX = C_U = \\ &\cong \frac{1}{N} \sum_{i=1}^{N} U_i U_i^{\top} \,. \end{aligned} \tag{D.54}$$

In this way, $I(\vartheta)$ can be calculated using only the gradient of Q; a similar reasoning applies for I_{te}. Naturally, the approximation (D.52) can be very rough, especially if P does not belong to the family $Q(\vartheta)$. However, it can be observed that $I(\vartheta)$, $I_{te}(\vartheta)$ always appear in the formulas multiplied by η, so the effect of this further approximation should be mitigated.

Bibliography

1. Aitken, A. C. (1935). On least squares and linear combinations of observations. *Proceedings of the Royal Society of Edinburgh, 55*, 42–48. Reprinted by Cambridge University Press. (2009).
2. Baarda, W. (1968). A testing procedure for use in geodetic networks. *Publications on Geodesy, New Series* (Vol. 2, No. 5). Netherlands Geodetic Commission.
3. Bengio, Y., Goodfellow, I. & Courville, A. (2017). *Deep learning* (Vol. 1). MIT press.
4. Betti, B., Crespi, M., Sansó, F. & Sguerso, D. (1995). Discriminant analysis to test non-nested hypotheses. *Geodetic theory today: Third Hotine-Marussi symposium on mathematical geodesy L'Aquila, Italy, May 30-June 3, 1994* (pp. 259–271). Springer Berlin Heidelberg.
5. Billingsley, B. (1995). *Probability and measure*. John Wiley & sons.
6. Box, G. E. P. & Tiao, G. C. (1992). *Bayesian Inference in Statistical Analysis*. Wiley.
7. Cox, D. R. & Hinkley, D. V. (1979). *Theoretical statistics*. Chapman and Hall.
8. Cramer, H. (1946). *Mathematical methods of statistics*. Princeton University Press.
9. Cybenko, G. (1989). Approximation by superpositions of a sigmoidal function. *Mathematics of Control, Signals, and Systems, 2*, 303–314.
10. Davidson, J. (1994/2003). *Stochastic limit theory: An introduction for econometricians Oxford, 1994* (Online ed.). Oxford Academic.
11. Dermanis, A., & Rummel, R. (2000). Data analysis methods in geodesy. In A. Dermanis, A. Gruen, & F. Sansó (Eds.), *Geomatic method for the analysis of data in the earth sciences*. Lecture notes in earth sciences (Vol. 95). Springer Berlin, Heidelberg.
12. Dirac, P. A. M. (1930). *The principles of quantum mechanics*. Oxford University Press.
13. Fieguth, P. (2010). Statistical image processing and multidimensional modeling. *Information science and statistics*. Springer.
14. Fisher, R. A. (1922). On the mathematical foundations of theoretical statistics. *Philosophical Transactions of the Royal Society of London A, 222*(594–604), 309–368.
15. Gelman, A., Carlin, J. B., Stern, H. S., & Rubin, D. B. (1995). *Bayesian data analysis* (1st ed.). Chapman and Hall/CRC.
16. Gikhman, I. I., & Skorokhod, A. V. (2004). *The theory of stochastic processes II*. Springer Science & Business Media.
17. Halmos, P. R. (1956). *Lectures on Ergodic theory*. AMS Chelsea Publishing Series (Vol. 142). Chelsea Scientific Books/American Mathematical Society.
18. Hansen, B. E. (2022). A modern Gauss–Markov theorem. *Econometrica, 90*(3).
19. Hastie, T., Tibshirani, R., & Friedman, J. (2013). *The elements of statistical learning: data mining, inference, and prediction*. Springer.

F. Sansò, A. Albertella, *The Probabilistic Vision of the Physical World*, Lecture Notes in Geosystems Mathematics and Computing,
https://doi.org/10.1007/978-3-031-88268-5

20. Heard, N. (2021). *An introduction to Bayesian inference, methods and computation*. Springer Nature.
21. Herbert, F. (1965). *Dune*. Chilton Books.
22. Hoerl, A. E., & Kennard, R. T. W. (1970). Ridge regression: Biased estimation for nonorthogonal problems. *Technometrics, 12*(1), 55–67.
23. Ito, K. (1954). Stationary random distributions *Memoirs of the College of Science, University of Kyoto. Series A Mathematics, 28*(3), 209–223.
24. Jeffreys, H. (1983). *Theory of probability*. International series of monographs on physics (3rd ed.). Clarendon Press.
25. Kirsch, A. (1996). *An introduction to the mathematical theory of inverse problems*. Springer.
26. Koch, K. R. (2007). *Introduction to Bayesian statistics*. Springer Berlin, Heidelberg.
27. Kolmogorov, A. N. (1950). *Foundations of the theory of probability*. Chelsea Publishing Company.
28. Kolmogorov, A. N. (1957). On the representation of continuous functions of many variables by superposition of continuous functions of one variable and addition. *Doklady Akademii Nauk SSSR, 144*, 679–681. *American Mathematical Society Translation, 28*, 55–59 (1963).
29. Lindley, D. V. (2000). The philosophy of statistics. *Journal of the Royal Statistical Society Series D: The Statistician, 49*(3), 293–337.
30. Lorenz, E. N. (1963). Deterministic nonperiodic flow. *Journal of Atmospheric Sciences, 20*(2), 130–141.
31. Lucchetti, R. (2006). *Convexity and well-posed problems*. CMS Books in Mathematics. Springer Science, Business Media Inc.
32. Lyapunov, A. M. (1992). *The general problem of the stability of motion*. Taylor & Francis.
33. Marzocchi, W. & Jordan, T. H. (2014). Testing for ontological errors in probabilistic forecasting models of natural systems. *Proceedings of the National Academy of Sciences, 111*(33), 11973–11978.
34. Mayo, D. G. (2011). Statistical science and philosophy of science: Where do/should they meet in 2011 (and beyond)? *RMM, 2*, 79–102.
35. Moritz, H. (1995). *Science, mind and the universe. An introduction to natural philosophy*. Wichmann.
36. NOAA National Centers for Environmental Information, Climate at a Glance (2024). *Global time series*. https://www.ncei.noaa.gov/access/monitoring/climate-at-a-glance/global/time-series.
37. Papoulis, A. (1977). *Signal analysis*. McGraw-Hill Book Company.
38. Planck, M. (1943). *Knowledge of the physical world*. Giulio Einaudi Editore.
39. Popper, K. (1962). *Conjectures and refutations: The growth of scientific knowledge*. Basic Books.
40. Popper, K. (2005). *The logic of scientific discovery*. Ed. Routledge.
41. Rohde, R., Muller, R., Jacobsen, R., Muller, E., Perlmutter, S., & Mosher, S. (2013). A new estimate of the average earth surface land temperature spanning 1753 to 2011. **Berkeley Earth Surface Temperature Project**. http://berkeleyearth.org/data/.
42. Sacerdote, F., & Sansó, F. (2005). Optimal linear estimation theory for continuous fields of observations. *Inverse Methods: Interdisciplinary Elements of Methodology, Computation, and Applications*, 262–275.
43. Sansó, F. (1986). Statistical methods in physical geodesy. In H. Suenkel (Ed.), *Mathematical and numerical techniques in physical geodesy*. Lecture notes in earth sciences (Vol. 7). Springer, Berlin, Heidelberg.
44. Sansó, F. (1997). *Quaderni di trattamento statistico dei dati; complementi di teoria della probabilitá*. Cittá Studi Edizioni.
45. Sansó, F., & Migliaccio, F. (2020). *Quantum measurement of gravity for geodesists and geophysicists*. Springer Geophysics.
46. Sansó, F., & Sampietro, D. (2022). *Analysis of the gravity field: Direct and inverse problems*. Springer Nature.

47. Sansó, F., & Sideris, M. G. (2013). *Geoid determination: Theory and methods*. Springer-Verlag.
48. Sansó, F., Betti B., & Albertella A. (2019). *Positioning, posizionamento classico e satellitare*. Cittá Studi Edizioni.
49. Snieder, R., & Trampert, J. (2000). Linear and nonlinear inverse problems. In A. Dermanis, A. Gruen, & F. Sansó (Eds.), *Geomatic method for the analysis of data in the earth sciences*. Lecture notes in earth sciences (Vol. 95). Springer, Berlin, Heidelberg.
50. Stocker, T. F. (2011). Introduction to climate modelling. Springer Science & Business Media.
51. Tanner, M. A. (1993). *Tools for statistical inference* (Vol. 3). Springer.
52. Tarantola, A. (2005). *Inverse problem theory and methods for model parameter estimation*. Society for Industrial and Applied Mathematics.
53. Teunissen, P. J., & Kleusberg, A. (Eds.). (2012). *GPS for geodesy*. Springer Science & Business Media.
54. Tikhonov, A.N. & Arsenin, V. Y. (1977). *Solutions of ill-posed problems*. Scripta series in mathematics. Winston & Sons.
55. Togliatti, G. (1976). *Fondamenti di statistica*. Hoepli.
56. Vapnik, V. (2006). *Estimation of dependences based on empirical data*. Springer Science & Business Media.
57. von Mises, R. (1964). *Mathematical theory of probability and statistics*. Academic Press.
58. Wackernagel, H. (2003). *Multivariate geostatistics: An introduction with applications* (3rd ed.). Springer.

The manufacturer's authorised representative in the EU is Springer Nature Customer Service Centre GmbH, Europaplatz 3, 69115 Heidelberg, Germany. If you have any concerns regarding our products, please contact ProductSafety@springernature.com

Printed and bound by CPI Group (UK) Ltd, Croydon, CR0 4YY
02/10/2026
02226883-0001